Donald 'Lofty' Large was born in 1930 and joined the Army as a boy. He fought in Ko   nent   ison-of-war-camp. After his re'   .tat.   AS and   rve in various conflicts around the world, from Malaya to Oman, from Aden to Indonesia. After his retirement, Large wrote two books about his Army career, preceding such authors as Andy McNab and Chris Ryan. He died in 2006.

www.transworldbooks.co.uk

SOLDIER AGAINST THE ODDS

FROM THE KOREAN WAR TO THE SAS

LOFTY LARGE

MAINSTREAM PUBLISHING

EDINBURGH AND LONDON

First published in Great Britain
in 1999 by
Mainstream Publishing
Transworld Publishers
61–63 Uxbridge Road
London W5 5SA
A Penguin Random House company
Mainstream paperback edition published 2000
Mainstream paperback edition reissued 2014

**SOLDIER AGAINST THE ODDS
A MAINSTREAM BOOK: 9781780576954**

Much of this text originally appeared in *One Man's War in Korea* (1988)
and *One Man's SAS* (1987) published by William Kimber & Co Ltd.

A CIP catalogue record for this book
is available from the British Library.

Addresses for Random House Group Ltd companies outside the UK
can be found at: www.randomhouse.co.uk
The Random House Group Ltd Reg. No. 954009

The Random House Group Limited supports the Forest Stewardship
Council® (FSC®), the leading international forest-certification organisation.
Our books carrying the FSC label are printed on FSC®-certified paper.
FSC is the only forest-certification scheme supported by the leading
environmental organisations, including Greenpeace. Our paper procurement
policy can be found at www.randomhouse.co.uk/environment

CONTENTS

This book is dedicated to all trogs. The few winners, the many losers and those, like me, surprised survivors.

To my wife, whose ever ready support through many trials and tribulations has made life worth living. And to the best parents a man could ask for.

To the special memory of Andy, Bernie C., Bernie W., Big Ron, Bob K., Bob T., Buddha, George, Jack, Kevin, Keith, Lofty, Lon, Nick, Nuge, Paddy, Titch, Tommy, Vic and Yogi.

To those who survived the war in Korea and in memory of those who didn't.

Ours – with cool calculation, sheer terror, screaming agony, savage anger and timely compassion – not to reason why?

INTRODUCTION BY ANDY MCNAB

Being like Lofty was something that I always aspired to without realising it.

When I joined the Regiment I was told that the best way to survive those first years in the Sabre squadron was to pick out someone who you thought you would like to be. Shut up, watch and listen. For me there were a number of the 'old and bold' who filled that requirement. It wasn't until later in my service that I learned that most of them, as newly 'badged' members to a squadron, had picked Lofty.

Joining the Regiment is about becoming part of a chain of élite soldiers through history, who have fought in every climate and terrain. Lofty had experienced more than most. His first operations were during the Korean War with the Gloucestershire Regiment. At the Battle of Imjin River – wearing a shirt and tie as part of his battledress – he was taken prisoner and held in a Chinese PoW camp. Lofty survived, despite a horrendous gunshot wound to his arm, even smoking marijuana to relieve the intense pain.

Once released, he didn't want to leave the army with a disability pension but stayed in and made a startling recovery to pass selection. He joined the Regiment for even more operations which took him to the jungles and deserts that were soon to become the Regiment's 'Operational Home'. He gave up the shirt and tie for jungle fatigues that rotted on his body as the jungle took its toll on the material during the long, hard and dangerous patrols that he had to endure. He was one of the men who developed the basis of the unique jungle skills that the Regiment uses today. He, and his fellow soldiers, did this the hard way, making mistakes that cost lives but learning from them so that future generations wouldn't die from those same mistakes. The soldiering he experienced is still of operational value today. It has passed down the generations. Long may it continue to do so.

Within the Regiment learning from other people's experience is all important. Even with the best training and hours of practice, everything can change in a moment on active service. The ability to adapt and to improvise is vital. Lofty was one of the first to understand this. His take was that 'Lady luck . . . is usually gracious enough to smile upon me when the shit hits the fan, so I have survived where better men have perished.' Lofty was a survivor, but it's not just about luck. He was too modest to say that it's also about guts, thinking faster than the enemy and being willing to go further than they will to survive. That takes courage and skill.

Not only was he a leader in the field but he was also one of the first to write books. Many people think that it has only been in recent years that non-commissioned officers of the Regiment have written their memoirs, but this is not the case. Lofty's *One Man's SAS* and *One Man's War in Korea*, published years before my first book came on the scene, were recommended reading for Regiment candidates. He was instrumental in setting the template for future members of the Regiment.

I have been privileged to meet some of the men mentioned in this book, and now I have met many others through its pages. Most of all, I am honoured to have known Lofty – with twenty-seven years in the British Army, he had 'Been there' and 'Done that'. He told it from a trog's – an ordinary soldier's – eye view, but let us into his thoughts and feelings as the action took place. Lofty didn't class himself as an exceptional trog, 'just an ordinary Brit whose experience was twisted this way and that by the unpredictable hand of fate'. We get to read about the official acts but also the unofficial, and how he came to play his part within them. There are many accounts of straight action on the bookshelves, but what I find even more interesting is seeing and hearing how an individual views that action and why he made the decisions he did. Lofty was critical where he believed it was justified, but still proud of what he was part of. I agree with his belief that 'any trog, in any army who is not proud of his unit, his army, his country, may have my sympathies but is not worth his salt'.

I hope that this revised and expanded single-volume edition of his two earlier books (long since out of print) will also make the list of recommended reading. It deserves to.

FOREWARNING

If you want a book by a professional writer, read no further; I was a professional soldier. Not the kind who sought high rank, but one who sought adventure and, always believing I was on the right side, the chance to do some good in the world while seeing a lot of it. The experience of achieving these modest aims is what this book is about.

My first, and worst, experience of war was with the Infantry in the Korean War. Since then my experience at the sharp end has been multiplied considerably with the SAS, in several areas of the world.

It is said that if you are wise you will learn from your experiences. Very true. In fact, it is good commonsense to learn most from the experiences of others. As a professional soldier, with the urge to go always a little further, yet live a little longer, other people's experiences – used in training – were of prime importance. Otherwise your own experience, being never enough, could be your last, or anything –ever.

Sometimes a lot of experience can be gained in a few short hours or days. It is more usual, however, for real experience to be gained over several years, in a variety of situations, different environments, different parts of the world and with a lot of luck. Not the luck which wins money in gambling or business, but that which ensures – when the smoke clears – you are the bastard still standing.

Due to the metal-wrapped high explosives thrown around with great force by various weapons on battlefields, and even in small skirmishes, combined with the comparative fragility of the human body, soldiers sometimes have to cope with sights beyond the comprehension of most people. And that is the last I will mention of such things in this book; I will leave the blood and guts to the writers of fiction, ghost writers and the like, who would need the ability of a writing genius and the disjointed brush of Picasso to get it even half right.

I have written about those things I have seen or done myself, except where stated. Many experiences have been left out for various reasons. Those of which I have written have been picked out because they were either typical of the situation or unusual in some aspect worthy of mention.

My twenty-seven years in the British Army were mostly enjoyable and unforgettable. They were made so, even under the most adverse conditions, by the men who served with me. I was fortunate to have always served alongside some of the best men on this earth. Those men I followed, and those who followed me, make me proud to have been a trog*.

---

* In the British Army: trog is an abbreviation of troglodyte. A term applied only to non-commissioned ranks of any army or real military force. Soldiers, like cave-dwellers, must sometimes live in holes in the ground to survive.

1 JAPAN—KOREA

The great white troopship, trembling with the power of its engines, rumbled steadily north-eastwards through the East China Sea, on its voyage to Kure, Japan. We watched the flying fish zipping and skipping in all directions as their peaceful world was disturbed and scattered by the bow wave.

His Majesty's Troopship, *Empire Orwell*, was on the last leg of her journey from Southampton, carrying reinforcements and replacements for the two British brigades fighting in Korea. Ninety-odd of us had joined the *Orwell* at Hong Kong, the last port of call, having all volunteered for service in Korea with the Gloucestershire Regiment. No doubt we each had our own reasons for volunteering, but I can only speak for myself.

I had joined the army in February 1946 as a fifteen-year-old band-boy. Before joining I knew practically nothing about the army, except that I had always wanted to be a soldier. During the Second World War I had seen quite a lot of soldiers, British and American, on field exercises around my home in the Cotswold Hills, so I thought that was what the army was like. Oh, the innocence of youth! Having lived some very formative years through the Second World War, it was obvious that soldiers in the British Army wore a khaki uniform, were shot at quite a bit but were allowed to shoot back. They also did their job in all kinds of interesting places around the world – that was about the limit of my real knowledge about the army. I had been in the Army Cadet Force and fired a few real weapons, but didn't even know I could have joined the real army at fourteen, straight from school.

It had to be the Infantry, to be a real soldier, and, as there were no vacancies for band-boys in the Glosters at the time, I joined the Wiltshire Regiment. I did not join the army to become a musician and soon hated music and everything to do with it; from the playing side that is. By good luck and a bit of fiddling – not of the musical kind – I was transferred to the Corps of Drums when we joined the battalion in Germany in 1948.

So, at seventeen and a half, I became a drummer, which opened the way to get clean away from the music scene. Several escape attempts during two years in Germany failed through accidents or sickness. Then, in 1950, we moved to Hong Kong and a few months later my great chance came to join the Glosters in Korea.

Luckily, in the Wiltshires' Corps of Drums, we were fairly well trained in weapons and field training, although most of the five years I spent with the Wilts was one long round of bullshit and drill. (In the army 'bullshit' is that which baffles logic either through actions such as polishing, painting, scrubbing, pressing and smartening to a point far in excess of being clean and tidy, or through talking a load of crap.)

I had realised while very young that a smart right turn would not dismiss a mortar shell any more than gleaming brasses would deflect a sniper's eye. But the emphasis placed upon these things in some quarters could be misleading. For my first five years in the army I was almost drowned in bullshit. I am no stranger to polishing in little circles on boots (even the bloody soles!), chin straps, white blanco on belts, straps, gauntlets, anklets, slings, drum ropes, leathers and almost anything which would hold still long enough. Cheese cutters (peaked caps) with and without slashed peaks, leopard skins and all the razzmatazz which was supposed to have impressed or frightened the Zulus, or someone, many years ago – and patently didn't.

I have marched out smartly as right marker for a battalion which was simmering in bullshit, without raising as much as a snarl from the RSM. So I have seen my share.

Before learning about bullshit and drill, while immersed in it, and for all the years since, I knew it was doing not one scrap of good for my ability to do a soldier's job. Even to a naive country bumpkin like me it was extremely frustrating to see the time wasted on drill parades which could have been spent on good, sound military training – at absolutely no extra cost to the army or the country. The equipment was there, the rooms or areas within the camps were there, the instructors, with vast experience gained in the Second World War were there (though getting steadily fewer). But the lack of brains, commonsense or whatever at the top, wasted men, time and obviously money on drill and bullshit. Thousands of good men left the army because of it. I wanted to be a real soldier, so had to get away from it all to get the training, the experience and do the job.

So, in early 1951, the thundering engines of the *Empire Orwell* were taking me and hundreds of others towards our first taste of war at the sharp end. The main event in my life so far, apart from joining the army, was meeting my future wife. Ann was working as a nanny at the Wiltshires' Depot in Devizes when the battalion returned from Germany, and we became good friends before the move to Hong Kong. She promised to write often – little realising what that would mean over the next two or three years. As the troop train pulled out of Devizes station I gave Ann my lucky mascot, a Cornish 'piskey' charm, given to me by a good friend from Cornwall years before. She still has it.

The trip from Hong Kong to Japan took about five or six days, at the end of which we found ourselves in JRHU (Japan Reinforcement and Holding Unit) on the outskirts of Kure. The camp at Kure was pretty basic but comfortable by 1951 standards. As it was a Commonwealth Forces Base, run by an Australian contingent, we were on Australian rations – the best rations I can ever remember having – even I didn't need double rations!

We were re-badged, re-equipped and brought up to scale for the colder climate of Korea, with heavy-duty pullovers, extra socks, heavy boots and a newer type of field equipment (1944 pattern).

There followed medical and dental examinations, carried out *en masse* in the barrack rooms, twenty-odd of us standing by our beds, starkers. More anti-this and anti-that injections then, after a few days, the ninety-odd ex-Wiltshires were sorted out into two groups. One group of about thirty, who were troops with battle experience, were sent off to Korea. The rest of us were marched out along country roads for twenty-odd miles to the battle camp at Hara-Mura.

I don't remember how long we were at Hara-Mura, two or three weeks I think, and it was well spent. We worked from dawn to dusk – often from dusk to dawn as well – marched miles and fired all sorts of weapons. We attacked and defended everything in sight, the right way and the wrong way to bring out the problems; became accustomed to the crack of bullets passing quite close. The battle school was run by a little, oldish major in the Somerset Light Infantry. He had under his command some of the best and most experienced instructors I have ever seen. They did an excellent job for which I have been forever thankful. There was absolutely no bullshit. We were expected to appear clean and tidy every time we turned out, night or day, rain or shine. Our huts were never given an inspection which we had time to prepare for. (There was little to do to the actual huts except sweep them out as required.) Staff would wander in at any time and expect to find everything clean and tidy.

Weapon cleaning was at an absolute premium, weapons were inspected at any time of the day or night without warning – anywhere we happened to be. We learnt to clean our weapons as soon as possible after firing or carrying them in the field. As we never moved around in trucks, always marched, we learnt to clean weapons on the march and during quick five-minute breaks at the roadside.

I wish I could remember that major's name. He was white-haired, about five-foot-three-inches tall, had three or four rows of campaign ribbons and was the first real officer I ever met. I have had occasion to be thankful to him many times since. Several times on exercises around Hara-Mura we stopped by farms or isolated houses and were often invited in for a drink of tea, or to have our water bottles replenished. We found the Japanese country folk to be very clean, hardworking and hospitable.

On one occasion I nearly came unstuck when an old farmer gave me a broad wink and poured a clear liquid into my empty water bottle. It looked like water but I was suspicious because he winked, and poured it like it was gold. It turned out to be very strong saki, Japanese rice wine. The way it tasted, the half-full water bottle was enough to sink an army. Anyway, we shared it around and the day felt very much better, much to the delight of the farmer and his family. Had I not noticed and tipped it down my throat like water, they would have had more to laugh about, no doubt.

At the end of the battle course we did a forced march back to JRHU Kure. There we were split into two groups of thirty. The second group would leave for Korea about two weeks after the first group. I was with the first group – which, within a few days, started on the journey to join our new battalion.

A troop train took us north from Kure to Sasebo, a port on Japan's northern coast. On the way we passed through Hiroshima, where the scars from the atomic bomb of nearly six years earlier were still visible, but rapidly being built over by man and grown over by nature. The staging camp at Sasebo was an all-American set-up; there were a lot of new things to see and get accustomed to, as we knew we could be involved with American forces in Korea.

Somewhere at Sasebo, perhaps at the entrance to the mess hall there was a great archway over the door, and with bold letters a foot high stating: 'Through these portals pass the world's finest fighting men'. We had never seen anything like it, and the comments drawn were many and varied. Such as, 'How did they know we were coming?'

There is supposedly an old Yorkshire saying, 'What's bummed-up needs bumming up!' – and that about summed up our feelings towards a lot of things we were to encounter from Sasebo onward.

I don't remember staying overnight at Sasebo, I think we went almost straight to the docks and the troopship – after a meal or two. Somewhere along the way we were called forward and checked off by American staff, probably just before boarding the ship, and a typing error really threw a lot of people into consternation. My army number is not really an army number but a regimental number of the Wiltshire Regiment. Instead of the usual eight digits, mine has only seven, so it's different to start with. I don't know how it happened, but instead of calling my number, rank and name, which was 5577627 Private D. Large, the big American military police sergeant yells out – after a long pause – SS.77627 Private D. Karger. As it came in the right place and I was expecting it, I recognised my last five numbers and answered readily enough. All stared hard as I walked forward and through a doorway. There was a lot of muttering in the background and later I noticed staff pointing me out to others who presumably had not been present at the roll-call. Whether or not any of the British Contingent were asked what they were doing with an SS man in the ranks, I never heard. No one asked me anything, in spite of a lot of curious stares. Perhaps being over six feet tall had something to do with that, though.

The ship we boarded was an old Japanese troopship. We boarded at night, slept our way over to Pusan, Korea, and were off the ship in the morning. All I remember of the ship was the decks, which were covered by thick straw matting on which we slept, and it was small for a troopship. We landed in Korea on 20 March 1951 – being greeted at the quayside and played off the ship by an American military band whose marching rendition of 'Saint Louis Blues' was brilliantly memorable. I wondered if it was a permanent job and how many times a day did they have to perform?

Korea, 'Land of the Morning Calm', was not enjoying many calm mornings in the early 1950s. In 1945, after the Second World War, Korea was divided roughly in half along the 38th Parallel. North Korea was administered by the Russians, so became a communist state; South Korea, administered by the Americans, stayed free. There were attempts by the United Nations to hold free elections throughout Korea so that the country could be brought together again under one government. Nothing could be agreed so Korea remained divided. The Russian and American troops were withdrawn in 1948. On 25 June 1950 the North Korean army crossed the 38th Parallel in a *blitzkrieg*-type invasion of South Korea.

American forces were deployed to help South Korea in July 1950, other United Nations troops arrived soon after. Nevertheless, the UN forces were pushed back until, early in August, they almost had their backs to the sea in what was called the Pusan Box. In September 1950 a UN offensive began, which practically destroyed the North Korean Army and UN forces gained ground almost to the Yalu River, the border between Korea and China.

At that point communist China entered the war and, catching UN forces at full stretch from their supply bases, threw in a massive offensive on 27

November 1950 which sent the UN reeling back far south beyond the 38th Parallel. By March 1951, UN forces had regained some ground and pushed the communists back to the general area of the 38th Parallel. The Land of the Morning Calm had seen some extremely bloody fighting in that nine months before I first set eyes upon it, and had been headline news almost constantly. In Pusan we again came under Commonwealth control, stayed one or two nights, then boarded a very rough and ready (maybe) troop-train for somewhere a long way north. We were given American rations to cook and eat on the train.

On the journey we were mixed with American troops, some were new replacements like ourselves, some were men returning to their units from leave in Japan. The countryside we passed through had been ravaged by war. By early 1951 a lot of Korea had seen the front line three or four times – and it showed. Nevertheless, people were tilling the fields. Occasionally some paused in their labours to wave at the passing train.

Towns and villages were in ruins, but when we slowed to a crawl – which was often in a built-up area – the people looked cheerful, considering their circumstances. Sometimes the train stopped for a few minutes and we would be quickly surrounded by a crowd, mostly of children, holding out their hands for the sweets and chocolates which showered from the carriage windows.

American rations seemed to us to contain more sweets and chocolate than real food, so the Korean children did very well for goodies when the Brits came through. At a small station somewhere well up the track the usual crowd of kids were being driven away from the train by American Military Police. This caused a lot of comment from the train, but one single act almost caused more than comment: there was a woman near the train, with a baby slung on her back in a blanket, the normal method of carrying small children in that part of the world. As the Military Police came along, she started to move away, but not quickly enough for one big sergeant who ran after her, repeatedly lashing her across the shoulders and back with a riding crop, the baby obviously catching a lot of the blows.

There was uproar on the train as Brits and Americans alike yelled their disapproval. A good friend of mine, Vic, leapt to his feet, slammed his bayonet on to his rifle and dived headlong out of the window. He didn't quite make it, I grabbed his belt and hung on. His rifle sling had caught on a metal fitting on the window frame and that held until help arrived, then it took three of us to keep Vic under control for the next few minutes until the train pulled out. Vic was a big lad, normally very quiet and well behaved, but right then he had only one thought, which he let everyone know: 'Let me stick this bastard up his arse and see how he likes it!'

Among the uproar on the train I could hear yells indicating that Vic was not on his own trying to leap forth into battle.

As far as I know, no one managed to get off, but I hate to think of the consequences if just one or two had broken loose. One American sat near me had pulled his .45 pistol and aimed it at the Military Police before being disarmed. But I didn't know this until afterwards when one of his friends returned it to him – with a lecture. His main excuse was, 'What the hell are we fighting for if that bastard is allowed to live?' He had a point.

We saw civilian trains sometimes, the ones going south were crowded with people hanging on the outside of the carriages and sitting on the roofs. At the side of the track every so often there were the bodies of civilians, men, women and children who had presumably fallen from the crowded trains. Eventually, somewhere, we left the train and were organised on to trucks, with all our kit, to take us to the first of a chain of echelons leading to the battalion at the front.

The first place we stopped was, I think, brigade rear echelon of 29 Independent Infantry Brigade, of which 1st Glosters were a part. A night or two there, then on the trucks and a long bumpy ride to the Glosters' rear echelon. Another night in a tent in a field, then on to the next place until we arrived at Battalion Headquarters of 1st Glosters.

We were checked, inspected and checked again. Given a list of kit and equipment we would require on the line, handing in the remainder to be sent back to one of the rear echelons. We were then sorted out according to rank, experience and specialist knowledge, and allocated to companies. Vic and I were to go to 'B' Company.

There was a pep talk by the adjutant, Captain Farrar-Hockley, and an interview with the commanding officer, Colonel Carne. The interviews were done individually in the CO's office, where Joe Carne shook hands and welcomed each man to the unit, asked a few questions – which showed he had seen our record documents – and wished us the best of luck for our future with the battalion. There were thirty of us. Joe Carne saw every man and made everyone feel welcome under his command.

The adjutant gave us a quick but plain briefing on the battalion positions and the situation along our part of the front. Then we were collected by representatives of our respective companies and trudged off to merge with the new environment.

It seemed about two miles to the company positions. We hoofed along winding, dusty, undulating vehicle tracks and narrow footpaths between the hills. The flattish valley bottoms were, or had been, cultivated, but the steep hillsides were covered in long grass, brown and dead from the recent winter. There were also a lot of shrubs and small trees dotted along the hillsides and on their crests, the crests being only a couple of hundred feet high – or even less – near us. In places the trees and shrubs closed up to form dense thickets. The scars of shells and bombs showed here and there. In some places they had caused small landslides which now showed as bare, light brown earth among the vegetation and outcrops of dark natural rock. In the distance we could see much higher hills and mountains far away.

Major Harding was company commander of 'B' Company. He welcomed the six or seven new faces to his company in a manner which made us feel really welcome, shaking hands with each one and putting friends together in the same platoons as far as possible. Vic and I were sent to No 6 Platoon where we were again made to feel at home. (Especially by the men we were replacing, who would be heading for the UK as soon as we were pulled out of the line.)

The company position was a group of small hills, each platoon had a patch of high ground to itself. As we climbed the narrow track to 6 Platoon's position a brew was put on, which was very welcome by the time we dropped our heavy kit.

We were sent to the same section, shown our trench, where to sleep, our arcs of fire and the breaks in the wire. More ammunition and grenades were issued, then we were briefed by the platoon commander, Lieutenant Peal. He pointed out the other company positions, the Imjin River and one of its crossing places. The hills to the north, beyond the river, were 'Indian country'.

From all the different briefings, and answers I got to a few questions put to the old hands I realised we were not expecting trouble in that sector. The battalion was due to pull back to a rest area within the next few days and, as a 'forty mile' penetration patrol with a lot of tanks and infantry a few days before had found no sign of enemy to the north, there seemed to be no problem.

Nevertheless, a brigade holding a divisional front was pretty thin on the ground. 'B' Company, as right wing of the battalion, had to put out a contact patrol every night to meet a similar patrol from the Northumberland Fusiliers. By this means they covered the two-mile gap which existed between the two battalions.

Even in the ignorance of youth I remember thinking it wouldn't be so bad against a highly mechanised road-bound army, but the Chinese masses of infantry could flood through the gaps at will. As far as I could see, advancing infantry could ignore our positions and leave us to die of embarrassment.

Also in the ignorance of youth, there is a great faith that those at the top know what they are doing and could of course plug the gaps at a moment's notice.

The British 29th Independent Infantry Brigade consisted of three infantry battalions which, in those days, were Glosters, Royal Ulster Rifles and Royal Northumberland Fusiliers. Artillery support was given by 45th Field Regiment, Royal Artillery, with their tried and trusted 25 pounders, and 170th Mortar Battery, RA, with 4.2-inch mortars. 8th Royal Irish Hussars were the brigade's armoured element, equipped with Centurion tanks, which were then the best tanks in the world.

Further artillery support was available from the huge American 'five-fives' a long way back. Air support was by the US Airforce. Helicopters were a thing of the future in forward areas, rarely seen anywhere, even well away from the front. In fact I saw no helicopters in Korea in 1951, although I have learned since that some of our wounded were evacuated by chopper.

A Belgian battalion was attached to 29 Brigade at the time, being the Belgian contribution to the United Nations' effort in Korea. The 29 Brigade front was about seven miles. Three battalions were stretched thinly along this front with vast gaps between battalions and smaller gaps between companies. The Glosters were on the left, covering the main supply route (a road running north to south) through the area. Royal Northumberland Fusiliers (RNF) were in the centre and the right wing was the Belgian 'Capitol' Battalion. The Royal Ulster Rifles (RUR) were in reserve, a few miles back from the front line.

Morale in the Glosters was high, the experienced troops who were the main part of the battalion exuded confidence which was, in turn, passed on to the younger troops such as myself and the young National Servicemen (conscripts), of whom there were about one hundred and forty in the Glosters at that time. Although the all-round situation didn't look good, to say the least, the troops had great confidence in the brigade and its commander, Brigadier Brodie. Above all, there was a confidence in themselves and the other units in the brigade, such as 8th Hussars and the artillery units, as well as complete faith in the ability of the other infantry battalions. The men in my platoon had, between them, seen action in North Africa, Italy, D-Day landings, Anzio landings, Burma and most of the bits in between.

When I mentioned the gaps around us to one old sweat he just grinned and said, 'Don't even think about it, kid, anything happens, Brodie will blot those gaps with Artillery, and the Rifles [RUR] will be in there like shit off a shovel to sort it out in no time.'

We two 'green 'uns' slept well that night. Except when we did our turn on stag (sentry duty), when we listened to the rumble of artillery to the south, the whistle of shells high overhead and watched the flicker of explosions in the hills to the north.

'Harassing fire,' someone explained. 'Just to keep them busy – in case they get any bright ideas.'

There was another rather worrying little detail I remember. Vic and I were allotted a slit trench, but it was full of empty ration cans and rubbish.

We had started to dig another hole for the rubbish when we were told not to bother. 'We'll be moving out in a day or two – and there is plenty of room in the next trench if anything happens.'

We checked on the next trench, only a few yards away, got to know its three occupants (including the section Bren-gunner) who gave us our new positions to try for size and shared a brew of coffee. The trench was rather large for three, being about four metres long and nearly one metre wide. It was deep enough for us two tall ones so there was no need for alterations.

And so to sleep – hoping the opposition didn't get any bright ideas, like returning that 'harassing fire' for instance.

2 IMJIN RIVER

It is worth mentioning at this stage that the modern concept of a soldier in battle, with steel helmet, camouflage smock and slacks, automatic rifle and grenades hanging out of everywhere, had no resemblance to the British Army in Korea.

For a start, only one man in the Glosters wore a steel helmet, don't know who he was, but I heard he existed.

We wore the common-or-garden dark blue beret as issued, with our Glosters' badges gleaming proudly. I say badges, because the Glosters wear two badges, one at the front and one (smaller) at the back. The back badge is worn as a battle honour from a war long ago, when the regiment had to stand back to back (surrounded as usual) in order to win the day. Looking at the history of the Glosters, one could get the impression that some situations are habit forming. The good old(?) battledress of the Second World War was our field uniform, a blouse and trousers with the added refinement of two buttons on the trousers to hold down the blouse at the back. Heavy boots and canvas anklets finished the picture.

I actually went into action in Korea wearing the above with a shirt and tie.

On the shoulders and sleeves of our battledress we wore our unit and formation insignia. The Glosters' title flashes were the bright red of the Infantry, with the name GLOUCESTERSHIRE in white letters. This was across the top of the sleeve in a slight arc. Below the unit title was, on the left sleeve, the Wessex Brigade sign as worn by the county regiments from the Wessex area of England. On the right sleeve was the formation sign which, at that time, was 29 Brigade's black, two-inch-wide square with a big white circle in it, known to everyone in Korea as the 'Frozen Arsehole'. For some reason I never did get my Frozen Arsehole sewn on.

Our field equipment was the relatively up-to-date 1944 pattern. (The bulk of the army in those days still used mostly 1937 pattern equipment with the odd bit of 1908 pattern.) The equipment consisted of two ammunition pouches supported by shoulder straps and belt and a small pack carried high on the back.

The dillies who designed British Army equipment up to that date could only envisage how smart it would look on parade. (Or possibly when standing back to back!) The ammunition pouches, when full, had the effect of keeping one's chest up off the ground at any time when it was the last thing you wanted, and as you dived to the ground the small pack had the knack of pushing your steel helmet over your eyes – or rolling it off into the distance. With no steel helmets, we didn't have that problem in Korea.

Our weapons were verging on the antique. The only worthwhile thing we had for the Korean War up to April 1951 was the good old Bren-gun, probably the best light machine-gun in the world. There was one in every section, so it worked out at one between every ten men. The officers had Sten-guns, rather utility, wartime ones which had been designed primarily to mass produce and use captured 9-mm ammunition, not much hitting power and useless at anything over about fifty yards away.

The 'troggery' were armed with old bolt-action rifles. Great for accuracy at long range – in the hands of this expert or that – at Bisley or some other comfortable Rifle Club on a summer's day in England. But to the trog facing a hail of automatic fire from short range, a bloody liability. At least, to be fair, they were better than spears – just.

The British Army had for many years been the victims of one of the greatest cons of all time. We were told time and time again that well-aimed single shots with bolt-action rifles would beat all comers. An enemy with automatic weapons could not sustain an attack because they would obviously run out of more ammunition than they could carry.

Before arriving in Korea we were told not to worry about mass attacks by the Chinese, as only one in ten had a rifle. This was basically true enough, but the other nine had tommy-guns, burp-guns or light machine-guns. (Brens, like ours, but a different calibre). The difference in fire power was such that I felt it was almost like trying to spit in someone's eye when he's spraying you with a fire hose.

On 22 April 1951, a few days after I arrived to join the Glosters, we heard reports of enemy troops on the north bank of the river. No one seemed to take this news as anything of note. Life went on as usual around the positions.

We cleaned and checked our weapons every day, sometimes several times if it was wet or dusty. On this particular day everyone was busy cleaning weapons before last light. We also checked and cleaned our grenades, primed all the grenades in the position and made sure everyone had his fair share of the ammunition which had just been issued.

All these activities, and many more, were carried out with an air of completely normal routine. Vic and I followed the lead of the experienced men around us. There was the usual soldier's talk, the odd joke, friends cursing one another in no uncertain manner, quiet laughter, the loud and lurid curse over some minor irritation – it was the normal British Army scene of quiet confidence. In retrospect, after many years on operations, I have the feeling that things would have been no different with the Glosters' trogs even had they known the odds which were about to be flung upon them.

In the evening, a couple of hours after darkness had hidden the hills and valleys around us, the sounds of battle came our way and some shelling of the river crossing took place. We watched the shells exploding and also watched the lazy curve of tracer bullets arcing this way and that as 'A' Company became entangled.

'A' Company were on a feature called Castle Hill, slightly nearer the river than we were and they were the left wing of the battalion. They were the first company position of the Glosters to feel the full weight of the Chinese attack. An ambush party, from 'C' Company, who were back in reserve near Battalion HQ, had made a good mess of the enemy attempts to ford Gloster Crossing, which was near the centre of the battalion front. But another Chinese force crossed further west and immediately hit 'A' Company.

Apart from the occasional long line of tracer which seemed to float gently towards us then pass by with a savage crackle, or thump into our hill, we seemed at first to be merely spectators at an average firework display. This was the first time I heard the Chinese bugles. They were very distant but also very creepy, haunting, eerie. Not the harsh, strong, brassy note of British Army bugles but sounding something between a hunting horn and a French horn. The notes rising and falling on the night air, competing with the sounds of battle.

The Chinese used their bugles to communicate orders and information, as the British Army had done years before. In this respect they were better off than us as our radio sets were very unreliable. Throughout the coming battle we would be hearing a lot more bugles, much nearer too.

Among the explosions of the grenades being used by both sides, we could hear the heavier thump of mortar bombs, but whether they were ours or Chinese was impossible to tell. Although I understood mortars and how they were used, and had fired our light mortar in practice during training, I could only imagine what it would be like to be on the wrong end of mortar fire.

Mortars were notorious casualty-makers during both World Wars. Being very light to carry, as well as very effective, they are the infantryman's artillery. Even our two-inch job had a range of five hundred metres.

Mortars fired at a very acute angle would send a bomb high in the air to fall – like a bomb from an aircraft – on to the target area. So the firer could be completely out of sight in a deep trench or behind a hill, and still give the target area a hard time. The bombs (sometimes called shells these days) explode on impact with almost anything, even small tree branches, sending blast and shrapnel in all directions. My lack of experience in that department would not be a long-term problem.

There was nothing we could do to help 'A' Company as they were about three thousand yards from us, well out of support range. 'D' Company was roughly halfway between 'A' Company and us, so, although still almost a mile from 'A' Company, they were able to give a small amount of machine-gun support until they themselves came under attack. The terrain, however, prevented us from supporting 'D' Company. Nevertheless, we were on full 'stand-to'. Everyone in his appointed place and ready for action. Well, fairly ready, that is. In my particular trench (next door to our real one) we decided to have a brew. This was a simple enough operation involving a tommy cooker, a mess can of water and a poncho over the top to conceal any light, all in the bottom of the trench.

There were five of us now in that slit trench, which was plenty big enough, but because of the brewing operation Vic was sat with only his feet dangling in the trench at one end and I was in a similar position at the other end. We were making sure no light showed, also keeping an eye on the slope of the hill.

The first explosion took us completely by surprise. We had seen or heard nothing. In retrospect this was unusual, as on all later occasions the Chinese advertised their presence with a lot of hawking and spitting as well as often chattering like a nest of magpies.

There had been no sound of incoming shells so I was pretty certain it was a grenade. There was no damage, but I found myself lying on top of everyone else in the trench, obviously being last in. Vic had a bit of help, as the explosion partly blew him into the trench. There were yells from the bottom of the trench as they were forced down upon the cooker and can of near boiling water.

Someone in a nearby slit trench yelled, 'Stop those grenades, you're too close to us!

Another explosion, closer. Then another, and I noticed something wrong. The British '36' Grenade has a noise all of its own. Those explosions didn't sound right if they were our grenades.

Being on top of the heap – and feeling decidedly exposed around the back end all at once – I looked over the forward parapet and to my surprise saw two or three figures just beyond the wire, about twenty-five yards away, calmly lobbing grenades at us.

The Bren-gun was in position on the parapet right in front of me so, without further thought, I grabbed the familiar weapon and plastered a long burst of about ten to fifteen rounds at the dim outlines I could see. A last explosion followed and I could see no sign of the men at the wire.

In the silence which followed I heard a voice from up the hill behind me shouting, 'Cease fire, stop firing that bloody Bren.'

I thought, 'All right for you, Jack, they weren't trying to blow your bloody arse off.'

Within a few seconds the company commander and company sergeant-major were at our trench demanding to know why the Bren had been fired. I explained and they looked doubtful, so I don't know where they thought those grenades were coming from. But I was a new replacement and likely to fire at shadows.

They were just about to leave when a hail of fire from a burp-gun chewed up the track just behind their crouching figures. All three of us immediately fired at the gun flash. Then there was silence.

The company commander said, 'Okay lad, keep your eyes peeled.' Then they were gone, silently into the darkness, up the hill.

All through this event, which probably lasted about three minutes from the first explosion, the brew-maker in the bottom of the trench had stuck to his post, and now produced two or three steaming mugs of tea or coffee which were shared around. It was when I took my first sip that I realised just how dry my throat was. I also realised that when firing the two or three shots at the muzzle flash of the burp-gun (with my rifle, as the Bren-gunner had taken over his own weapon), I had done something by instinct rather than training. I had used the rifle like a shotgun and kept both eyes open, not trying to use the sights (which were useless anyway in the darkness).

Having used shotguns from a very early age, it was, I suppose, quite normal to revert to habit when surprised at close quarters. Nevertheless, as I quietly sipped my hot drink there was the worry that I had betrayed my training, not 'done things right' in the first operational use of my rifle. Little did I realise I would spend a lot of time, many years later, being trained in exactly that type of instinctive shooting.

No. 6 Platoon was not bothered again that night, except by stray bursts of heavy machine-gun fire from a long way off. I believe one or two of the other platoons in 'B' Company had probes like ours, but nothing serious.

The experienced troops were more interested in their tea and coffee than anything else and I've seen more flap on training exercises than I saw that night.

The sounds of battle rose and fell in the distance as attacks went in on one or another of the other company positions. Flares hung in the sky, slowly drifting on their parachutes. Signal flares arced this way and that. The flashes of exploding grenades and mortars were an almost continuous flicker. Tracer was everywhere and, although often covered by louder explosions and firing, the steady 'tac-tac-tac' of long stringy bursts from the Glosters' Vickers medium machine-guns seemed to stitch it all together.

Much to my dismay, there were no bodies lying where they should have been when daylight came. It looked like I had fired at nothing, but an old hand found tracks and marks and explained how the Chinese always recovered their men, dead or wounded, in the night when possible. As would we in a similar situation.

Soon after dawn, a solitary Chinese soldier came jogging along the valley in front of our position. The company commander gave permission and our Bren-gunner fired a short burst which killed the enemy soldier. Three or four of us then went down the hill to collect any documents from the body. I kept a sharp lookout for any Chinese dead further down the slope but saw none. We did see a neat row of six Chinese back packs not far from where the enemy soldier was lying, but did not touch them.

The fighting in the distance became sporadic as the sun rose, the war seemed to be going around us.

We were quite surprised when orders came to pull out immediately, leave everything except our fighting equipment and small packs and move south, up a long spur to higher ground. In fact, Colonel Carne had by then assessed what we were up against and was trying to organise his battalion into a smaller defensive area so that we could still deny the enemy use of the main track, and also prevent companies being overrun one at a time without a chance of support from other companies.

The company moved quickly. A few old hands rapidly turned the position into a death trap, with booby traps of all kinds, as soon as we were clear. Our heavy greatcoats were thrown in a heap and I saw grenades being placed in them soon afterwards.

We also dumped our picks and shovels and that left us rather exposed on our next position. Even with the proper digging equipment, I don't think we would have made much impression on the solid rock I found in the area of my next dig. In retrospect, I believe the original intention was to fall back from one hill to another until the battalion reached hills a few miles back where, a couple of weeks earlier, it had prepared defensive positions (for just this eventuality). But things changed when we were ordered to hold fast before we reached the prepared defence line.

Some reports, read over the years since the battle, claim that the 'typical British Army understatement' used by our brigade commander, Brigadier

Tom Brodie, on the telephone to the American general commanding that part of the front was misleading. He is said to have told higher command that 'things are a bit sticky'. Whereas, in American terms – and in reality – we were in deep shit. But the situation of 29 Brigade was just not realised by those in command until far too late. The Americans, of course, had plenty of problems of their own at the time with communist forces attacking their front in great numbers.

We moved off in a long column with 4 Platoon leading, followed by 6 Platoon, then Company HQ followed by 5 Platoon.

The company was about six hundred yards from its old position when the Chinese started getting among the booby traps. Several of them were gathered around the pile of greatcoats and I saw two or three on the ground after the first grenade blew. They went back to the pile and got caught again before I lost sight of the position.

The spur we were climbing was not too steep, but it was long. Before we had gone far I noticed my friend, the Bren-gunner, with his extra load, was beginning to blow a bit, so I offered to carry the Bren for him. He gladly accepted.

Of course, being young, fit and about six feet six inches, the Bren was no trouble to me, whereas the Bren-gunner was a bit below average height and quite a bit older, with a couple of rows of war ribbons. I also took from him a small haversack full of loaded magazines and the spare-parts wallet. It's no use carrying a machine-gun in action if you can't reload it or clear any stoppage. The spare barrel remained with the Bren number two, as normal. He was also carrying a fair number of magazines. Besides the machine-gun magazines, we all were carrying several bandoliers of ammunition, each holding fifty rounds in clips of five. The rifles and machine-guns used the same ammunition: .303 inch, rimmed.

We were right out in the open, on a long grassy stretch, when I noticed a US Air Force Shooting Star ground-attack fighter going around us in a wide circle. There was a sudden mad scramble to lay out the brightly coloured recognition panels. This was done only just in time. The jet fighter was just stacking up for his attack run when he must have seen the panels. We had no ground-to-air radio communication at company level and those panels were our only hope.

Before leaving our area, the Shooting Star slaughtered a lot of Korean civilians who were running south in the valley at the side of our spur, the aircraft passing so close that we could see into the cock-pit and watch the pilot, intent on his job.

I felt a surge of anger at this seemingly wanton attack on innocent civilians, and no doubt made some comment. One of the old hands turned to me and said, 'It makes everyone sick but you have to get used to it. The North Koreans have used civvy clothes to infiltrate our lines before now – so nobody takes any chances.'

In the comparative silence which followed the departure of the jet, the wailing of a baby from among the mess in the valley did little to convince me there wasn't a better way to fight a war.

We were nearly at the top of the hill into a fairly thick cover of small oak, ash and undergrowth, when all hell broke loose up front. Within seconds 4 Platoon came racing back through us, by which time our platoon had fanned out in an extended line and hit the deck, weapons ready.

I still had the Bren-gun, so looked for the Bren-gunner. Not being able to see him I ran forward to a position where I could see the top of the hill, hit the ground and prepared to give covering fire as required.

The platoon sergeant was nearby and said: 'What the hell are you doing?' I told him what I thought I was doing and he said: 'Well, get your backside ready to move fast. We want all the Brens in the assault line.'

As he said it Lieutenant Peal, 6 Platoon Commander, yelled: '6 Platoon – on your feet – fix bayonets.' There followed the clatter of bayonets all around in the trees, which was probably heard by the enemy above us.

I thought: 'Shit! This is for real!' And the adrenaline pump went into overdrive.

Then the command: 'Advance!' And off we went, up the hill in an extended line abreast. I knew the Platoon was about thirty-strong but, because of the trees and curve of the hill, I could only see three or four other men to each side of me. We were aware of each other but our attention was focused on the top of the steep slope.

As we moved forward, Lieutenant Peal told everyone to watch the top for grenades and, as soon as they appeared, run like hell under them 'and get among the bastards'. The Glosters had obviously done it this way before, as everyone seemed to know what to do.

I watched the skyline above me and, sure enough, when we were about twenty-five yards from the top a flock of hand grenades came flying to meet us. I ran up there as I had never run before.

The grenades exploded behind me and I went over the crest, almost running into the enemy grenade throwers, who were completely taken by surprise. There were four or five of them in various positions of shock on the ground, mostly to my right. The Bren-gun raked along them then I almost ran over one right in front of me. He had a stick grenade in his right hand. He was almost lying down on his back but supported by his right elbow, his left hand stretched toward me, the fingers splayed out as if to ward off what he knew must come, his tommy-gun was lying across his lap.

Had this been the top of the hill I may have been able to take him prisoner, he was young, probably about my own age, and in absolute shock. But it was a false crest and the hill was now a gentle slope, the real crest hidden by trees and a thicket of undergrowth.

This was the first enemy soldier I had ever met face to face in broad daylight, and we were almost close enough to shake hands. The machine-gun roared its short burst of two or three rounds, but there was no elation as I jumped over the thrashing body, whose face and expression will never be forgotten, and rushed on, firing from the hip at any movement I saw. I found myself entering a much thicker clump of trees with a lot of undergrowth and bushes. There was a considerable amount of wild privet or something such and I remember thinking it would be difficult to hold a straight assault line, which is essential if you are not going to shoot one another. That was the last time during the assault that I was aware of other members of the Platoon, who were rushing forward to my left and right. I pushed forward as fast as I could to make sure none of them got in front of me. A few grenades were exploding but I don't know whose they were, and there was a continuous uproar of small-arms fire. There was screaming coming from somewhere nearby, but no time to do more than notice it.

Thanking my lucky stars I had copied the Bren-gunner's example and placed two magazines inside the front of my jacket, I went down on one knee for about two seconds and changed to a full magazine after firing about twenty-five rounds, then rushed on. Its no use having a lot of ammunition if you can't get at it quickly. (My ammunition pouches at that time contained some spare socks, some brew kit, a few other odds and ends and two or three grenades.) Once moving again, I unclipped the top of the magazine haversack, not knowing how many more I may need. The Bren-gun is a beautiful instrument for inspiring confidence (when you are on the right end of it!). Its thunderous fire went before me like a shield of death. A burst of automatic fire came from almost in front of me. Two or three short bursts with the Bren-gun either closed him down or frightened him off, and that was the only return fire I was aware of. Glimpses of mustard-coloured uniforms among the trees and the half-seen silhouettes of two running figures attracted a burst or two each from the smoking machine-gun. Any extra thick undergrowth collected a raking burst – just for luck. Whether or not I hit any enemy at that stage I've no idea. The trees and undergrowth prevented any clear shots, so I just sprayed every movement and kept moving forward. Anything suspicious on the ground or through the branches collected a couple of rounds before I passed it, whether or not it was moving. Some ammunition was wasted on what turned out to be discarded enemy packs. All resistance seemed to crumble before us, until I reached the other side of the position, having (luckily) just changed the magazine and only fired a couple of short bursts from the new magazine.

About ten yards in front of me was a Chinese soldier, we saw each other at the same moment and both fired at the same time. He was standing on a narrow ledge, just below the edge of a cliff, so that all I could see of him was from the thighs up. He had a tommy-gun and really let it rip. Most of his fire went over my left shoulder, all of mine ploughed up the ground in front of him. At ten yards' range we emptied our weapons and both missed. I would have said it was impossible.

He jumped sideways and back over the cliff as I did a fast reload and rushed forward. By the time I could see him again he was picking himself up from a slide of shale and began to run. I yelled to him to stop as I knew he had no chance (and I was pretty green anyway). Of course he didn't stop, so I kept firing until he dropped.

We had taken the hill in a few short minutes and, once organised, found we had no real casualties, only a few slight wounds between the two leading platoons.

In military terms, 'B' Company had successfully withdrawn about a mile and taken Hill 314. As someone in 'B' Company remarked at the time, 'It felt like we were advancing in the wrong direction.'

I had my cap comforter (a woollen hat, used mostly on night patrols) rolled up and tucked under the shoulder strap on my left shoulder. Amazed to find it full of holes, I realised the tommy-gunner had not missed by much. Luckily, only one bullet had hit the rolled-up hat, but it had gone through many folds.

'B' Company's air of quiet confidence prevailed. The hill position was taken and secured in text-book manner – as if it was just another exercise on a well-trampled training area. There was no apparent excitement at our small success, just a business-like determination to do the job and prepare for whatever might be our next trick.

Personally, I was full of suppressed emotion. Pride at being part of a unit which could go in with such determined and obviously unexpected speed. Relief that – on my first real assault – I had done most things right, not felt much fear, hit what I had tried to hit (mostly) and got away with it unscathed. My excitement of youth was, however, soon dampened by the task of searching enemy dead for documents. Dead men are no problem, but they have photos in their pockets which give them reality. Little black-and-white snapshots of smiling wives, children, parents and grandparents. Happy groups whose smiles will be wiped away by the war. Those good folk had done nothing to deserve this. Ours not to reason why.

The Company took up defensive positions on the hilltop, but having only one or two shovels in the Company (captured when we took the position) and the hill being mostly rocks, we were unable to dig in properly. All we could do was work with our hands, bits of wood and bayonets to scrape away the shallow soil and push most of it to the forward edge of the position to form a low breastwork. The position was quite good for cover against attacking infantry, as it was at the top of a steep slope and the thicker trees and undergrowth were all behind us, giving us a good, fairly clear field of fire downhill.

We were right at the edge of the wood, with branches spreading above us. If only we could have dug in properly. As it was we had practically no protection from the back or sides, so the thoughts of incoming machine-gun or mortar fire were not for dwelling on. Also, if assaulting troops reached the top of the slope to left or right of us, we would be almost completely exposed to their fire. Nevertheless, morale was high, in spite of the shortcomings of our situation, and we were confident we could hold our ground when the time came.

We had a good view of the slopes to our right and could see about thirty yards to our left, so could give some support to other positions.

My particular arcs of fire were left and front. The Bren-gunner would be able to move according to circumstances, but was mainly right and front. His number two, the third man in the position, would keep him in touch with the best, most needy, machine-gun targets. All very tidy, and organised – until the shit hit the fan.

Vic and three others went off with a stretcher, carrying the last Chinese soldier I had dropped, and who, riddled with bullets, was amazingly still alive. He died just after reaching the regimental aid post. His comrades were all either dead or had managed to escape the assault.

About that time, as we were preparing to repel the inevitable attack, we heard that 'A' Company had taken heavy casualties and lost their company commander, as well as most of the other officers.

All around us the battle was hotting up. Artillery fire and air strikes were loud and long to the north and west. The haunting, eerie notes of the Chinese bugles were now much nearer and clearer, mostly to the north and west. I wondered if the foxes back home felt this way as the huntsman's horn sounded steadily nearer and nearer.

We had just got down to solid rock, about six inches deep, in our fire position when someone called me to the top of the hill, gave me a pair of binoculars and told me to let the company commander know when I could see the American Infantry on a ridge to the east. I was told they were coming in to help us out and should be there soon.

In fact, the ridge to the east of the Company position was part of a mountainous hill called Kamak-san which rises to two thousand feet and, although none of the Glosters battle touched it, it had dire consequences for the battalion. After we withdrew from our original positions in the low hills near the river, Kamak-san became a wedge between us and the rest of 29 Brigade. It prevented any support reaching us from the east and channelled the Chinese 63rd Assault Army on to the Glosters position. It was too distant from the roads and tracks in the area to be of any use to anyone, so was left undefended and unattacked, standing in rugged isolation as the tides of war swept by on either side.

Through binoculars I could see the river and the flat area to the south of the crossing. The Chinese were taking heavy casualties in that area, where they were wide open to air attack and well-directed artillery fire. In spite of my, situation I felt sorry for those poor bastards who, like me, were just men doing their job.

We could hear the artillery firing somewhere to the south and, in the comparative quiet of what was a fine spring day, the whistle of shells passing high overhead. Then the great flash and sudden appearance of a fragmented, dark, cloud of smoke and dust among the lines of tiny figures running south from the river. After a few shells the whole scene became hazy, partly blotted out by smoke and dust.

Then the artillery stopped. The scene cleared. Lines of men were still running south towards the battalion positions.

I wondered what had happened, then heard the whistle and howl of the diving jet fighters. Two of them screamed across the sky, low over the massed humanity like rigid silver birds of death. I didn't see the napalm bombs leave the aircraft, but I did see them burst.

A hell of burning petroleum jelly, like a line of fire from a volcano (from my viewpoint), spread through the still-running lines amid a great cloud of black smoke. I watched in fascination as men ran flaming from bursts of napalm to stop and fall in the long grass, leaving a smouldering trail and a little bonfire where they fell. The Shooting Stars climbed away with a thunderous roar of their engines and stacked up for their next run, this time with rockets. We could hear the tearing rush of the rockets then the sharp crash of explosion as they struck somewhere among that inferno of smoke and flames.

Again the silver birds climbed across the sky and again they swooped on their prey. This time the crackling roar of cannonfire followed immediately by the deeper roar of the exploding cannon shells.

By then the aircraft were hidden from us at their lowest point by the clouds of smoke and dust over their target.

A couple of runs with cannon shells then, presumably out of ammunition, the US Airforce went off for their coffee and doughnuts. The whistle and thunder of their engines faded into the distance, and we heard the rumble of artillery somewhere to the south again.

Those readers unfamiliar with the ins and outs of military tactics, abilities and logistics may wonder why the Chinese attacked us on our hill positions; why they didn't just ignore us and push on with their advance to the south, why they had to buy those hills at such a high cost in the lives of some of their best assault troops?

It was not the hills they wanted, nor was it directly the destruction of the battalion. What the Chinese had to do (if their great offensive was to succeed) was eliminate the observers who brought in the air strikes and artillery fire. Their supply lines could not operate until those people were silenced, and without adequate supplies and reinforcements their offensive was doomed.

The foregoing does not mean the Glosters themselves were not causing enough problems. Their mortars and machine-guns were effectively brought to bear on any enemy road movement within reach, but there had to be a limit to their ammunition once they were cut off from the main UN forces. The air attacks and artillery barrages, on the other hand, were more devastating and not subject to the same limitations.

While on watch on the hilltop someone called our attention to a section of Chinese who were walking along a track to the east of our hill. The track was on the side of the next hill and, when first seen the enemy section, nine tommy-gunners and one Bren-gunner, were walking south, almost towards us. The track could be seen curving east opposite our position then it disappeared into trees to the east. At its nearest point it was about three hundred and fifty metres from us and much lower.

The Company Commander gave permission for riflemen to fire at this tempting target, but no automatic weapons were to be used. About five of us, those who had a clear view of the enemy, commenced firing when the enemy section had almost reached its nearest point.

At the first shots the enemy stopped. One or two went down on one knee, most just stood still.

Then, to my great amazement, they continued on their way, almost ignoring us. I say 'almost' because, apart from the initial pause, the only reaction we got was from the leading man who, at the nearest point to our position, carelessly pointed his Thompson in our general direction and squeezed off a short burst of three or four rounds – without pausing in his steady walk.

The most pointed and dramatic 'Up yours, Jack!' I've ever seen.

The army classed me as a pretty good shot. I readjusted my rifle sights and tried again; I tried bloody hard. Not one of the enemy was hit. I suppose we should all have been grateful to the last man of that enemy section for not raising two fingers in time-honoured fashion before he disappeared into the trees about four hundred metres away.

An old soldier sitting nearby, who had not fired a shot said, 'Let that be a lesson to you, your shooting got worse every time you missed. Think about it.'

I thought about it, long and hard, to realise I had been beaten by psychology. The enemy had known, as I know now, that if they don't get you with the first shots – when they think they have everything right – then it's unlikely the shooting will improve. If you are lucky enough not to be hit by the first few shots it is most likely the fire will become very inaccurate and demoralise the opposition. A nil reaction to enemy fire makes it very ineffective, a fact which I have put to good use on several occasions since.

It was at about that time that I realised we were not up against a screaming mob who relied just on weight of numbers, but were facing some very experienced troops. A lot of them were 'good old boys' who had been around a bit. The Glosters didn't have a monopoly on battle experience in that neck of the woods; a very sobering thought when I again turned the binoculars on the mess near the river and paid more attention to the hundreds who were escaping the napalm, rockets, cannon fire and artillery shells.

Later that day we came under a fair amount of sniper fire. By then I was back in my fire position, there having been no sign of any help coming our way. Someone said that the Yanks hadn't been able to break through.

I remember thinking at the time that if there was still only 29 Brigade on a divisional front against this lot, we were going to have a ball breaking out if the rest of the UN couldn't break in.

During that afternoon of 23 April, I watched a little bit of Brit 'hard neck' which takes some beating.

One of our platoon was digging a hole with one of the few shovels, right on top of the hill. He was getting steadily sniped at as he worked and, to my amazement, was using the shovel to indicate the snipers fall of shot. Whether or not he was telling the truth I never knew, but at each shot he would stop work, indicate an inner or outer point with the shovel to where the shot had passed him – then carry on with his work until the next round cracked past.

Night came and 'B' Company was still, I believe, reasonably intact. Before midnight the platoon sergeant took a couple of us down on a listening patrol to the saddle east of our position. We crept down the hillside, facing a full moon, towards a dark line of trees at the saddle. I didn't feel kindly towards the moon that night.

The steep slope was covered in dead grass a foot or so deep, with a few low bushes and the odd rocky outcrop or tree here and there. At any moment I expected enemy fire. They were all around, and I felt sure there would be some watching this area, waiting to assault the hill. After about one hundred yards the dark line of the trees loomed close then, with great relief, we moved into the deep shadows and stood silently for a few seconds to listen for any sounds of the opposition. The sounds of battle were a continuous uproar to the west, but we were descending the east side of our hill and the sounds became muffled as we went lower.

There was a sound, a swishing sound, almost like distant running water, but definitely not running water, which I could not identify, though it was alien to the woods at night. We moved on through the belt of trees towards the saddle.

As we reached level ground I could see through the lacework of branches to the track beyond. Trying to focus my eyes through the intervening branches I saw what I took to be a field-gun with armoured shield, wheels and trail standing on the track. As I contemplated what I saw, two huge men appeared. One grabbed the muzzle of the gun, the other grabbed the trail, picked it up and went off out of sight.

It took me several seconds to realise that what I had seen was just two ordinary Chinese soldiers picking up one of their wheeled medium or heavy machine-guns, which are more or less the same outline as a field-gun, but, of course, only a tenth the size.

The sound I had heard earlier was now louder but its source was only too apparent. It was the sound of the soft hockey boots of many men moving quickly past our positions under cover of darkness. It looked like we might yet die of embarrassment.

In less than a minute we had seen and heard enough and, to my great relief, started back up the hill. Within a few minutes of our arrival back in the company position our report brought down a hell of artillery shells on the column of enemy troops moving past to the east of us.

Sometime towards morning the mortaring started. I had been asleep for what seemed only a few minutes so had lost track of time. The first explosion awoke me and I could hear the swish of more coming. The next few hours were a hell of noise, screams, shouted orders, crackling machine-gun bullets, then scrambling, panting men coming up the slope spraying at us with their submachine-guns and falling in doll-like heaps as we fired and fired and fired again, the weapons bucking and pushing into our shoulders as they became hot and smoking. The haze of smoke, dust and dirt made our targets difficult to see sometimes. In the bright moonlight our attackers were, in my case, silhouetted against a light background of moonlight on the sandy earth and rocks at the bottom of the slope behind them. Even so I was often firing at the muzzle flash of their weapons.

From time to time, flares drifted in the sky over the hilltop, and 'varey' illuminating flares arced above us, as one side or the other tried to light up the scene. The effect, on this picture from hell in front of me, was to fill it with ever moving shadows, of trees, branches and men. As men were lit up then turned black by passing shadows, often their own, they seemed to appear or disappear in the rifle sights. Then, when the overhead lighting stopped there would be a terrible blackness, punctuated by muzzle flashes and grenade flashes, but leaving us feeling blinded and exposed to the now unseen.

There were a few scattered trees and rocks on the slope, so some of the enemy took cover there hoping to make the most of a steady clear shot or well-aimed grenade. We, however, had other ideas and watched for any sign of these lurkers. The answer to them was a No. 80 White Phosphorous grenade. The No. 36 grenade (our most common grenade) was a high-explosive shrapnel job with segmentation for fragmentation – or what we trogs called 'fair shares for all'. But anyone who was out of sight of the explosion, behind a log or rock, was usually pretty safe. The No. 80 WP on the other hand was totally different. Upon exploding it threw a mass of burning phosphorous all around which came down everywhere like falling rain. Without good overhead cover there was no hiding place.

My companions were experienced men but I thought they were joking when one of them said, as he prepared to throw the first 80 grenade, 'Keep your head down until the glare has gone then shoot the moving lights.' Sure enough it worked, though there was only one screaming lurker to shoot. On another occasion there were three or four. After that they went off lurking quite a bit.

Most of their dead and wounded finished up well away, down at the bottom of the steep slope, but a few were lodged nearer to us. The stench of what was smouldering after the 80 grenades will not be forgotten. The smoke from all this also gave them a screen to cover their next assault. I believe they also used it to remove some of their wounded. Most of the attackers would have been severely wounded rather than killed instantly by our defence, such is the reality of the infantryman's war at the sharp end, where killing is the intent, but disablement and a lingering death in agony has to suffice in most cases. There are no clean Hollywood battlefields. Even so, many who are severely wounded are not in great pain, shock is the greatest killer.

After the first hour or two the Chinese assault troops must have been climbing over their own dead and wounded down there below us.

Only one, as far as I know, reached the top of the slope in front of me. I'd had to reload the rifle, and looked up, just as the clip went in, to see him looming over me, almost on the parapet. I did the only thing I could and lunged with the bayonet. He did nothing to defend himself, just stood there. It's likely he was already dead before the bayonet hit him.

A lot of fuss and bolloxology has been made by various people at various times about the act of taking human life. In battle there is no thought, no hesitation, no remorse, no pity. Only mixed sensations of anger, fear, excitement, exultation all vying for pole position as you try to hit the right targets, like the one who is firing at you before the one who is nearer but stumbling forward like a man in a dream. The one in the act of throwing a grenade, before the one who is nearer but fumbling to pull the grenade pin.

There is also a terrible determination that they will not beat us, they will not move us – we will gladly kill the bloody lot of them.

The most dominant, overriding sensation in battle can probably be summed up in one word: apprehension – for your own skin, no one else's. When it seems certain you will be hit you are inwardly flinching from the unknown.

So much for the mythical sanctity of human life on the battlefield, or anywhere else come to that.

I have often wondered how many of the good, kind and gentle mothers of Dresden, Cologne, Leningrad and London would, given the means, have gladly blown seven shades of shit out of those brave young lads in the sky above them.

Among the hectic sensations of battle there was also a grudging admiration for the Chinese troops who seemed to exist by the maxim of 'try, try again'.

Not having air support and, at that time, no artillery to compare with ours the Chinese relied not only on a vast numerical superiority of infantry, but on the guts and go of the men in their infantry units.

I personally saw nothing for China to be ashamed of during the Imjin River battle, especially when considering these men had probably done a fast approach march of twenty or thirty miles with full battle kit and attacked straight off the march.

The bugles were very close now but often their strange notes were lost in the welter of sound all around us, only to come through again after a few seconds, clear and close – but still haunting and eerie.

At some stage I realised my right cheek was bruised and aching from firing the rifle which, when it kicked, slammed my thumb against my face. This was due to the rapid rate of fire and continual changing of position and point of aim.

Counting the shots: one, two, three, four, five, six, seven, eight – then worry about reloading another five or emptying the rifle. The panic, if I did empty the rifle, to ram in another clip and start counting again. At every opportunity I placed several prepared clips on a piece of brown paper in front of me ready to hand for quick reloading. The rifle would take ten rounds, there were five rounds in each clip.

I noticed the brown paper suddenly had a lot of blood on it and the clips were sticky when I reloaded but I have no idea where the blood came from. It wasn't mine.

There wasn't time to worry further than that, but there was time to worry that the sticky clips might jam the rifle.

The Bren-gunner seemed to spend most of his firing time either standing in the middle of the position or kneeling against the breastwork and firing the machine-gun like a rifle, so he could get the best view down the steep slope. Several times I was very much aware of the muzzle blast as he fired over me, then there was the flickering flash, lighting me up from above as the machine-gun roared back at the hail of metal coming the other way. Even now, in modern times, there are few machine-guns to beat the Bren in its adaptability to circumstances of battle, especially in expert hands such as this one was.

At the hint of a lull, when we had no one coming at us, there was the frantic activity of reloading the Bren magazines, pushing single rounds into the rifle to make sure it was full, checking grenades were primed and ready, to hand.

The two old hands with me produced their oil bottles and put spots of oil on the working parts of their weapons. I followed their example, trying hard to remember my training, anxious to get everything right now it really counted. These slight lulls in the general uproar did not, however, always mean the enemy was giving us (and himself) a break. When things went comparatively quiet for a few minutes we had to keep a sharp watch on the slopes below us, as it often meant a stealthier attack would be mounted, with the enemy crawling and wriggling up the steep hillside.

The mortars were getting air bursts in the trees and sending down a hail of shrapnel, branches and shards of wood. They also hit the ground and those sweating, struggling bodies lying in the open upon it. 'B' Company was being slaughtered, but not in spirit.

An obviously officer's voice shouted: 'Stand fast the Glosters – remember the Back Badge!' and a trog's voice shouted back 'F*** the Back Badge – I want out!'

'I want out!' was one of the current army sayings, voiced by all and sundry when things got too sticky, boring or bullshitting, and meant out of the army rather than the immediate situation. It went along with such sayings as, 'Roll on death. Demob's a bloody failure!'

I was amazed to find myself laughing and heard the laughter of others through the clatter and roar of the battle.

At times I thought there were only the three of us still alive. All the men near us were dead or horribly mutilated, no longer able to take part in anything.

The explosions around us, when close, seemed to knock or suck the breath from our lungs, and once or twice the blast threw me sideways or smashed down with pressure from above. Things sometimes became pretty hazy and blurred. Our eyes stung from the smoke and fumes which also made us cough and gasp for air. My ears hurt and I tried a few times to stuff dead leaves in them, then found moss and dirt was better but still gave very little relief. Then there was the worry that I would not hear the things I needed to hear.

My fingertips and knuckles were raw and bleeding from trying to scratch every last bit of dirt from the solid rock beneath me. Every tiny bit lower I could get, and every scrap of dirt I could push up around me, became all important in the struggle to survive. Digging my fingers into the cool soil – literally hanging on to Mother Earth – to hold on to reality in a world being shredded around me.

When someone came crawling and scrambling to us from behind, I thought we were getting some help. No such luck, they wanted the Bren-gunner to go somewhere else to fill a gap in the defences.

So there were two of us now and no Bren-gun. The screams of some of the men lying around us, friend and foe, were terrible, but there was nothing we could do except try to exist.

The next position to us, on our right, just a few yards away, also started with three men, but no Bren-gun. One man was killed early on, then a mortar shell or grenade took out the other two. One killed outright, I think, the other badly wounded was pulled back out of sight, probably by the medics. They also brought us a lot of spare ammunition and grenades from that position.

The position on our left was obliterated at some stage but, as it was on higher ground, we only noticed when we saw one of them dead, having been blown over the edge of the bank towards us. The grenades, mortar shells and machine-gun fire did not differentiate between the dead, the dying and the damned.

Just as first light was beginning to creep through the treetops someone brought us more ammunition and a lot of grenades. My companion and I worked as fast as we could to prime the grenades for throwing. A dozen grenades were still in their box and it was as much as I could do to keep my mouth shut when he started pulling pins from the primed grenades and placing them back in their pigeon holes in the box. He saw my amazement and explained, 'For emergencies. When it gets too hot we throw the bloody box and all.'

We had plenty of grenades and the way things were shaping up it looked like a good idea, but I was a bit worried the box might get hit or knocked over before we could use it.

A machine-gunner about two hundred yards away was picking on me every time I raised up to fire at the troops below. It was daylight, I took out my cap badge, then took off my beret, but still that bastard picked on me. He chewed the rocks to bits and blasted gravel into my face.

Unlike the enemy assault troops, who were completely exposed to our fire at very close range as they scrambled up the steep slope and had practically no target to aim at themselves, the machine-gunner was in a good position well hidden, somewhere on a spur of the next hill, north-west of our position, and every time we positioned ourselves to fire at the enemy below we were probably exposed to the waist in the machine-gun sights. Each time we fired at the men below I fully expected to be killed. The weight of fire from their automatic weapons, mostly tommy-guns and burp-guns literally chewed everything to pieces all around us. Earth, rocks, roots, grass, branches and twigs shattered in a crazy uproar of crackling bullets, screaming ricochets and shrapnel.

We must have been getting help from other 'B' Company positions, but their fire could not be noticed in the welter of enemy fire coming at us. The law of averages was against us. There would be ten or fifteen automatic weapons spraying their contents at us every time we showed – plus that damned machine-gun. There was probably a lot of other small arms firing at us from the cover of the trees and rocks from which the enemy mounted their attacks, and back to which their wounded crawled, rolled, dragged themselves or were pulled. That was about sixty or seventy metres away, but below us.

'B' Company was being steadily chipped away. Every assault caused a few more casualties but, every time they came at us there seemed to be more of them. More machine-guns joined in the uproar raking the hill top, there seemed to be more grenades exploding closer in front of us and to the sides and instead of the original one or two assault lines coming up the slope there were now three or four at a time. Not a great mass, but line after line, wave after wave breaking against us. In spite of the casualties we inflicted, they seemed to be multiplying in their numbers. According to the records, read years later, 'B' Company survived seven major assaults through the night and early morning of 24 April. There were also many lesser assaults no doubt, but none of us had time to count at the time.

Those moments, and many others since, when I threw my life to the winds of chance, come most clearly to mind upon noticing my army pension is taxed as 'unearned income'! I would love the person who dreamt up that one to stand where any of 29 Brigade stood that day so he could see just how to unearn income!

Another rush was coming up and I fired from a low position at those along the slope which I could see without rising up, but I knew they were coming up right at me from below as well

We knew the assault troops were very close when the mortar shells stopped exploding and the machine-gun fire stopped or was lifted to pass overhead. It was all a matter of timing. Come up too soon and you were badly exposed to heavy machine-gun fire. Come up too late and you wouldn't come up again. The old soldiers with me had, initially, been telling me when to do what, but it is something you catch onto very quickly – if you live long enough.

I threw a grenade just over the edge and fired a couple more shots at a few of them who were still coming further along. Then the grenade blew and I jumped up in a rage to nail anyone below me and that bloody machine-gunner as well – if I could.

He was a better judge of time than I was, a cooler head, more patient – and he nailed me.

The half-raised rifle was slammed back into me and a sledge hammer with a knife on the edge crashed me back in a heap.

I could hardly move. A great numbness engulfed my left side. The rifle was lying near me, its butt smashed away and the magazine and trigger destroyed. Expecting a tommy-gunner on top of me any moment, I found a grenade with my right hand, then, as my left hand wouldn't move, I jammed the rifle between my legs, hooked the grenade pin over the muzzle and pulled it free.

A few feet away the other man in our position was still firing as fast as he could, not realising he was now virtually on his own.

No tommy-gunner came to finish me off, instead, there came a sort of lull. That bloody machine-gunner was still blasting away every now and then as if to say, 'I'm still here', but there was a definite slackening of pace for some reason.

In the uproar of sound all around us I was occasionally aware of our own artillery or mortars firing very close support. I know that, at one stage, shells from the 25-pounders of 45 Field Regiment were falling among the enemy just in front of some of our positions. So close in fact that I worried about being hit by our own people.

I think it was the 4.2-inch mortars of 170th Mortar Battery which caused the slackening of pace to my immediate front. Every now and then mortar shells would land among the jumble of rocks, bushes and trees from where the Chinese were mounting their attacks. At first I thought the enemy had mistakenly hit their own men, then one of the people in the position at the time said, 'Thank Christ for the Royal Artillery!'

Right then I was ready to thank anyone or anything for a bit of a break and, agnostic that I am, envied those who could believe in something I couldn't – and even get some small comfort from that belief. Nevertheless, thanks to the Royal Artillery or some even higher power, we had a slackening of pace. About then I suddenly felt quite cheerful in the knowledge I had a 'Blighty'. (A wound bad enough to get me sent home – back to Blighty.) How the hell I thought I'd get out of there, I've no idea.

The other man looked around, relief on his blood-speckled face, as he realised we had survived another rush. Then he saw the blood and I said, 'I'm hit.' He stared in disbelief for a few seconds, then crawled under a poncho, and just shook.

A few minutes later Lieutenant Peal came over with more ammo, saw I was hit and asked where the other chap was. I looked at the poncho, Lieutenant Peal looked at the poncho. How he knew in that place of dead and half dead, I'll never know. He called the man's name, pulled back the bolt of his Sten-gun, and said 'You have three seconds to get out of there . . . one!'

The man came out like a startled rabbit, stared at Lieutenant Peal, then at me for a few seconds. Lieutenant Peal said, 'You okay now?' The man nodded and resumed his firing position, then carried on as normal as if nothing had happened, and was a damn good soldier too.

Remembering my training, I realised I should have taken out my field dressing and placed it on my chest, in text-book manner. But I had a live grenade in my hand with the pin out and couldn't move the other hand, so I lay there in absolute confusion, trying to decide what to do.

The machine-guns increased their crackle, but the mortars stopped and I knew another rush was coming, so my problem was solved.

I watched the only other living being I could see, and beyond him the slope further along. Then when he raised up to throw a grenade and I could see the mustard-coloured uniforms on the slope again I slung the grenade back over my head and with great relief fumbled for the field dressing. Somehow, the fact that I was likely to be killed in any number of messy ways any second, was as nothing compared with the importance of putting that dressing on my chest as per the book.

Soon after getting the field dressing organised, I had a nerve-racking experience. My left arm and hand would not move. I had tried hard –

who wouldn't in those circumstances? All one side of my body was numb and felt dead.

Suddenly I became aware of something moving at my side, glancing sideways, I saw an arm raising itself up with the fingers opening and closing. In my dazed state, I didn't at first recognise my own arm. When I did, I tried to control it, but couldn't. At one stage I thought it would claw me, but after a while it fell to my side again.

For some reason, I don't have any idea why, many of us that day had our sleeves pushed up to the elbow and our bare forearms were covered in blood – as were our faces – from scratches, pinpricks of rock splinters, bullet and bomb splinters.

Also we were dirty from scratching into the ground with our bare hands, and from blackened rifle oil. My arm didn't look like mine.

For what seemed a long time, I lay on my back, gazing at everything and seeing nothing.

The only things I remember were watching the toothpaste and shaving cream splashing out of my small pack as that bloody machine-gun hammered again, and even in the din all around I could occasionally hear the enamel popping from my army mug, still attached to a strap on the pack. The pack was about two feet from my head.

My companion was bobbing up and down, trying to get a look at the steep slope. Suddenly he turned, grabbed the box of grenades with both hands, shook it over the edge then threw it with a pushing motion.

I don't know if it was one of those grenades or one thrown by the opposition but something exploded right on the parapet, covering us with more smoke, dust and rubbish. He shouted something at me, then jumped to the parapet and fired several shots down the slope.

The look on his face when he got down again told me we would be okay from that direction, for a few minutes at least.

I think it was about that time that I noticed a couple of men with a Bren-gun were in or near the position to our right, which had been destroyed earlier.

It was a great comfort to me to see and hear a Bren being used professionally, with short sharp bursts which I knew would be accurate and most damaging to the enemy.

Lieutenant Peal came back and stood over me. 'How are the legs? Do you think you can walk?' I had no idea, I felt completely numb – but I nodded. The machine-gunner went berserk, the twigs and branches around Lieutenant Peal were alive with bullets, my mug spun madly on its strap again, dirt and stones flew from the bank.

Lieutenant Peal looked in the direction of the machine-gunner, squeezed off a short burst of three rounds from his Sten-gun in the same direction, said, 'I'll send a medic – if there is one,' turned and walked away, completely unscratched.

A few minutes later a medic came, cut the clothing around my left shoulder and did a quick job with a field dressing. Not mine, it had disappeared – that bloody machine-gunner perhaps.

The medic checked me over for more holes, but didn't find any worth patching, said, 'If you find any more holes that are leaking much, give us a shout,' and ducked away. He had dragged me back from my fire position, and a few minutes later a couple of men came, dragged me further back and I told them I could stand up. To my surprise I stood up and could walk. We moved a few yards into the trees and the next thing I remember is sitting with my back to a tree, staring at an arm and hand lying on the ground in front of me. There were several bodies nearby, at least one was Chinese. I wondered why the arm had been amputated, it didn't look too bad. Why I thought of amputation in that hell where I had already seen much worse, I don't know.

The two men came back, crouching to keep low. 'The walking wounded are being evacuated – come on.' They helped me through a mass of tangled, fallen trees to the edge of the cliff and seemed to be looking for a way to get down. It was the place I had reached when we assaulted the hill the day before, coming face to face with the tommy-gunner.

As we paused at the edge, a machine-gun tore apart a small sapling right by me. For the first time I knew real terror. The cliff was about fifteen or twenty feet and sheer to sloping shale. I just jumped straight out in absolute panic. Landing on my feet, I ran down the slope. Ahead of me was an archway formed between two clumps of alder bushes with an inviting dark thicket beyond. I realised I was on exactly the same route the tommy-gunner had taken the previous morning. All I could think was: 'History repeats itself.'

In blind panic I reached the archway and threw myself through, tripped and fell headlong down the hillside. My head hit something and I went out like a light. Regaining consciousness, I found myself draped over a fallen tree, all panic was gone. I immediately took stock of my situation. My left arm was still useless, hanging like a rope of pain and swinging around freely with every move. The whole of my left side felt on fire. Blood ran from a cut on my head down my face, getting in my left eye and dripping from my nose. Everything else seemed to be in working order.

Getting my bearings, I looked across the valley to the east, where the twenty-odd walking wounded were supposed to be going. Seeing some of them going into a small valley about two hundred yards away, I knew I had not been unconscious for long.

Before I could move, however, I became aware of a face, staring at me from ground level, about twenty yards down the hill. It was a Chinese face, so I stood still and stared back at the glittering dark eyes. There was no movement and no gun in sight. I glanced left and right, but could see no signs of other Chinese. So it was just him and me.

We stared at each other for several seconds, then I realised something was wrong. He hadn't moved or blinked. Nothing. A few seconds more and I walked down the hill and past him. He was very dead. A victim of the artillery barrage of the previous night. Then I ran. I passed several more dead Chinese, through shell craters, tangled branches and across the valley to where I had last seen the line of wounded. They were moving east, up a small valley, climbing directly away from our company position which towered behind us.

I reached the end of the column as the men at the front turned south to climb out of the valley. They had picked a spot where there had been a landslide, where progress would not be impeded by the scrub and small cliffs.

As the leading men neared the top of the slide a machine-gun opened up from near our company position, which sounded like it was being overrun at last.

Men on the slope began to fall. The rest of us rushed on, trying to reach the top where we would be over the crest and into cover of trees. A man in front of me fell, then tried to get up again. Without thinking, I put my good arm around him and helped him to his feet then, still supporting him, struggled on up the slope. The slope was soft sand and loose rock. It was like the nightmare of trying to run in treacle. Two steps forward and one back.

More men went down, the bullets crackled around us. The man I was supporting was hit again and all the strength went from him. I could no longer support the dead-weight and, as he looked dead this time anyway, I let him drop and scrambled on up. He had shielded me from the bullet which finished him. To this day I am not sure if the instinct for self-preservation made me pick up a 'shield', or did I help him because he was there? The machine-gun fire was from my right, and only my right arm could support him; nevertheless, he stopped a bullet which would have hit me, maybe more than one bullet. As it was, I survived, yet again, and reached temporary safety over the top of the ridge.

At this point in my life I met one of the greatest men I have ever known. Over six feet tall, very broad shouldered, experienced soldier and always cheerful: Corporal George Baker from London. Like me, he was from 'B' Company, 1st Glosters. I had probably seen him around but not really met him. He was the man in charge of the walking wounded, charged with the task of getting us out through the Chinese lines if at all possible. George had gone back down the slope and helped others reach the top. He was not wounded and had a rifle, which gave me some hope.

There were now eight of us, including George Baker and a lance corporal who was not wounded, helping George get us out. The lance corporal had a Sten-gun with one bullet in the breech and the magazine housing was smashed. It was doubtful if it would fire even the one bullet. George had only two bullets in his rifle. We had no other weapons or ammunition between us.

The first thing George did was check the state of the wounded men. One or two extra wounds were dressed roughly. Someone made a sling for my arm with a face veil (small camouflage net), then we set off down the slope to the south.

We had no maps, no compass and no idea of the possible whereabouts of other forces – friend or foe. The only map I had seen was a very confused-looking thing which Lieutenant Peal had used on his initial briefing. He had stubbed a finger at the various battalion positions and the nearest company position of our neighbours, the RNFs. The only thing I had been able to identify was the Imjin River. We knew the Chinese could be up to twenty miles behind our lines, in great numbers. The artillery were still firing somewhere to the south but it was difficult to judge the range as the hills and cliffs threw echoes in all directions.

I personally didn't have a clue about anything and had not been in the country long enough to know anything about it. Although I had recently travelled up the road to the battalion area, 'yer average trog' doesn't see a lot to aid navigation when looking through a cloud of dust from the back of a truck.

We were all in the same boat, so it looked best to stick together, head south and hope to get through somehow. I knew which way was which and, with my natural built-in know-how, could have found Battalion HQ easily enough, but when last seen it had been in a valley, on the road, and there was no way it could have survived that position, so it would have moved or been destroyed. I also knew I would be of no use to Battalion HQ in my state of health so, altogether, there was no point in trying to find it.

Our instructions were to get out, 'out' was south, so south we went. After a while of walking and trotting we came to a road. Just a few yards to our left, to the south, there was a sharp bend in the road and the first two men set off round the bend to check it out. A few seconds later they came galloping back, yelling: 'Tommy-gun Platoon.' Then, as if to leave no doubt, a hail of automatic fire sprayed the corner as they left it.

Everyone bolted north up the road, but I stood in some indecision, as I thought it best to go back into the trees, away from the road. The herd instinct took over and I overtook everyone else before they reached the next corner.

The road was in a narrow valley at this point, with cliffs at the sides in places. We turned west off the road into the first valley we came to and ran for some way up a steepish track. After covering some distance up the track we noticed a long line of troops trotting along the ridge to our right. There was some discussion as to whether they were friend or foe. They were too far away to be sure, but I decided in my mind they had to be Chinese – there were too many of them to be anything else. Then we noticed they were turning south around the head of our valley so, as most thought for sure they were Chinese, there was no way out that way. They *were* Chinese, as we found out later.

The tommy-gun artists were still following us, the odd burst of fire told us that, so George decided to climb the ridge to our left, which was south anyway. The side of the ridge went up in steps. A steep climb, then a small cliff of rock, then another steep climb and so on. Some of the cliffs were only three or four feet, others were six or eight feet.

How we got up that hill I'll never know. George Baker was literally heaving the wounded up the cliffs. Not so bad with some, but I was over fifteen stone. At one place George pulled me up by my collar and belt while I scrabbled for a grip with my good hand.

At last we reached the top and the shelter of a young fir plantation. The trees were thickly planted, forming a dense mass, except at the bottom where there was a gap of about three feet between the lower branches and the ground.

The tommy-gun crowd were still persisting. We crawled and dragged ourselves under the trees for about one hundred yards, then they ended. Ahead of us was open ground, a shallow valley. On the other side of the valley there were scores of Chinese troops, digging positions and erecting camouflage. They were about one hundred yards from us. The tommy-gun crowd were firing the odd burst into the plantation behind us. It sounded like they had spread and were sweeping through.

There was a narrow track running down the valley between us and the Chinese positions. The track was only about thirty yards from us. Along the track there came trotting one little Chinese soldier, carrying a long old rifle and bayonet over his shoulder, clutching a piece of paper in his hand – obviously a messenger hurrying about his business. George Baker gave his rifle to someone able to fire it and came out with the quickest briefing – certainly the plainest – I've ever heard. He said, 'I'll go out and put my hands up. If he takes me prisoner, come out one at a time with your hands up, if he doesn't, then shoot the bastard and from here on we'll throw rocks.'

With that he stepped out of cover and raised his hands, leaving at least one of us – with great faith in human nature – feeling around in the pine needles for a rock.

The little Chinese looked startled, as well he might, with George up the slope from him looking like a 'man mountain' – only rougher. He glanced across at all the help he could get from a few yards away, took the rifle from his shoulder – but didn't point it rudely – grinned and beckoned George down toward the track. The little man's amazement grew as, one by one, we followed, until he had eight comparatively huge enemy soldiers all to himself – his prisoners!

He lined us up and marched us off down the track in his original direction. We had only made a few yards when the first of the tommy-gun crowd broke out of the trees, pointing their weapons and shouting at our captor, who, to my amazement, pointed his old rifle at them and really put them in the picture. I didn't need to understand Chinese to know that! Then off we went again, with our unwounded hands on our heads and our hearts deep in our boots.

3 INTO CAPTIVITY

Being captured by enemy troops – made a prisoner-of-war – surrendering one's life to the enemy, is the most degrading thing I can imagine. All men being different, there is no doubt that being captured had different effects on different men. I can only speak for myself, state the facts as I saw them. Nevertheless, I have good reason to believe most of us who were captured at that time – 22–27 April 1951 – had very similar feelings and reactions. A deep conviction of having failed. Failed our country, our unit, our family, ourselves – any one of those, or all. In my case, all. Why should I be alive and snivelling to the enemy for my miserable life when so many had died? What right had I to live when others had died for nothing?

The Chinese had obviously advanced far beyond our miserable little hills and we had failed to stop them. Was I a coward? If I had done this differently, or that faster, or something else another way, would it have made any difference? To anything? Or was I really surrendering to man's age-old instinct to survive? Undoubtedly braver men than me were captured that day.

I went over and over the events leading up to my capture. Should I have gone my own way when we first hit the road? Should I have tried to hide in the fir plantation? Neither had seemed a good idea at the time, so why should they now? They didn't.

The Chinese troops were absolutely saturating the whole area. Acres of them sat on vast hillsides, each man about ten feet from the next, holding a small branch over his head. They were probably invisible from the air.

Soon after capture we were handed over to a proper escort, at some kind of headquarters under some trees. Looking along the hillside, which was fairly smooth and open I could see for about six hundred yards (over half a kilometre) to where there were more trees. The whole open area was covered in troops just sitting there, with their little branches. They were spaced apart so as to minimise the effect of shelling or air strikes. They were far less likely to be attacked than were the obvious hiding places such as woods and ravines. Aircraft and artillery were likely to blast woods, because they afforded cover to the enemy, whether or not any actual target could be seen.

The Chinese uniform was much lighter in colour than ours. It was a light mustard colour when new but, as it was washed, became lighter and some were almost white. So, when sitting in the open with a little local camouflage to break the outline, there was no dark, eye-catching shadow.

Their weapons interested me a great deal. There seemed to be about half with tommy-guns (the genuine US-made model) or burp-guns (a Russian submachine-gun, the PPSh 1941) and the other half had a mixture of rifles. A rifle with folding bayonet seemed most common. This was the Russian Simenov.

The burp-gun was called so because of its very rapid rate of fire (nine hundred rpm). A couple of quick burps and that was seventy rounds gone – it was empty. Many of the tommy-guns and almost all the burp-guns were fitted with drum magazines, which held fifty rounds on the tommy-gun and seventy-one rounds on the burp-gun. Spare magazines were carried in good leather or canvas pouches at the waist and also, occasionally, in small haversacks slung over one shoulder.

Every Chinese soldier had four or five grenades hanging on the back of his belt. These were what we called stick grenades. They had a handle about seven or eight inches long. Their effect was (thankfully) more blast than shrapnel.

Most of their weapons had a plug in the muzzle, like a cork in a child's pop-gun, with a brightly coloured feather in it and a string attaching it to the weapon. I wondered if this was left in when the weapon was fired and just popped out, as per pop-gun, or if the brightly coloured feather was there to remind them to remove it before firing.

Most of their Bren guns were carried in leather cases, the shape of the gun, magazine and all, with a zip fastener. The magazines on their Brens were straight and probably only held twenty rounds.

On their feet some Chinese wore soft canvas and rubber hockey boots, others wore plimsolls, giving them the ability to move around quickly and noiselessly at night, whereas our great, heavy, steel-shod boots were a liability on night patrols, especially in the rocky terrain of Korea.

Spare ammunition was carried on the belt and in a bandolier over one shoulder. Over the other shoulder they had a cloth tube, about two or three inches in diameter, which contained their field rations. Their field rations looked and tasted like chicken meal. Adding a little water, they mixed it to a brown paste and ate it cold, just like that! So simple, not like our flashy rations, but so very effective in the field. The Chinese had been at war for a long time by 1951.

The Chinese troops were as curious about our equipment as we were about theirs. We were, of course, still enemies but we were all soldiers and understood and accepted the situation as real soldiers seem to do anywhere. There was no hatred between us. They had won, we had lost, but all were still soldiers.

Front-line troops of any army can be ruthless in battle – and usually have to be out of necessity – but once it is over the average soldier accepts a new set of rules immediately. They knew we would escape if the opportunity arose; we knew they would shoot us for trying if the opportunity arose. And that is about as far as a soldier thinks, on either side. There are usually more important things to occupy the mind in a battle area.

All races are different, but those who have nothing to prove rarely try to take it out on helpless captives or civilians – as far as front-line troops are concerned, anyway. In the rear areas of any army, well away from any action, there are often the glory boys, who have never, and are never (if they can help it) going to hear a shot fired in anger. They are almost always a different kettle of fish – usually gutless.

Generally speaking, from long experience, the more gutless the man – or the race – the worse they treat those who have no defence. There are exceptions, but these are usually due to religion or indoctrination.

At the Headquarters we were searched thoroughly, but nothing was taken from us, neither documents or valuables. Our treatment had, so far, been fair. We had been given cigarettes (Senior Service) and chocolate (Cadbury's Dairy Milk), obviously captured. I think I had two squares of chocolate and was thankful; it was the first thing I had eaten for nearly two days. Little did I realise it would be a long time before I ate again.

We were moved a couple of miles with our new escort and taken to a small valley, where we found about thirty other British prisoners were being held under guard. These Brits were mostly Glosters, but one or two other units were also represented.

While we were in that valley, we were shelled by our own artillery. Only four or five shells landed near us, but I noticed no one took any action. At least, most of us just sat or lay where we were. Only two men dived for cover. One was killed, the other was severely wounded and died later. They were in among the rest of us, it was as if the shrapnel sought out anyone who moved.

At that stage in the game we were bomb happy, had seen so much it didn't matter any more. We didn't give a damn for bombs or shells. They got you or they didn't. Rumour had it the Chinese had been doping their assault troops. For a start, I'm quite sure there wouldn't have been enough dope to go around, but there's no doubt most of them were bomb happy.

There is another phenomenon that affects troops in battle. For the first couple of hours of being shot at, you feel everyone is shooting at just you. After that first two hours (give or take a bit) if no one has hit you, then you get the idea no one can. This is not to say you think you are bullet-proof, but you have the confidence to take chances, can afford the luxury of more emotions than pure fright. Real laughter in the face of adversity can do wonders for morale; losing your temper can get your arse blown off!

More prisoners were brought into the valley. They arrived in small batches of sometimes four or five, sometimes ten or fifteen, until there were nearly one hundred of us there. By this time my wounds were a mass of agony, and things are again a bit confused as regards time of day, or even day.

While still in that small valley, a Chinese unit of about fifty tommy-gunners passed through. They were very fresh-looking, nothing like the battle-hardened veterans we had seen so far. Their uniforms were newer and cleaner, they were all loud talk and menace. We called them 'tommy-gunners' but, in fact, I noticed these troops were carrying what looked to be brand new burp-guns. The woodwork on their weapons was a highly polished, reddish-brown colour, and all had long curved box magazines. This was the Chinese version of the burp-gun, known as the Type 50.

We didn't know it then, but it is most likely these were some of the forward assault troops of the Chinese 65th Army which, in terms of manpower, was a second wave of twenty-seven thousand men being thrown at the rather battered 29 Brigade. They went through the group of prisoners and took watches, rings and anything they could tear loose. This was no more than we expected, but came as a surprise after the firm but comparatively courteous treatment we had received so far.

I was left alone, having no rings and with my watch being hidden by the makeshift sling. There was a fair bit of caked blood around and I don't think these darlings wanted to soil their sweet little hands. Our guards were obviously terrified of them. We had only about five or six guards at that time, and about forty or fifty of us.

Our guards stood well away from us, spaced in a circle on the slopes around us, so they could look down on the whole group. One had a Bren-gun, the others had Thompsons. As more prisoners arrived the guards were increased. By the time we moved out of there, at dusk, 24 April 1951, there were about thirty guards to about ninety or one hundred prisoners.

Just before we moved out of that valley we were addressed by an English-speaking Chinese officer who wanted drivers for the few trucks which had been captured, on the pretext that they would be used to transport our seriously wounded comrades. Sufficient volunteers were found and went off under guard.

A few weeks later the drivers rejoined us. They told how they had been conned and forced to drive the vehicles loaded with supplies. They never saw any of our wounded. Although they did not stay together as a group during their driving for the enemy, they all managed to sabotage their vehicles within a day or two, once having found it was a trick.

I don't remember much about that night, except pain, and marching this way and that. There was a thick fog, or river mist, and at one stage we got lost from those ahead. We were climbing over paddy field walls, banks of earth, sometimes four or five feet high and I had some difficulty. All I could think was: 'Keep going.' Eventually we located the rest of the column and at dawn were nearing the river. Joining a rough road leading towards the river we met thousands of Chinese troops running south. All had their trousers hanging around their necks. A column of 'flashers'. They had obviously just waded the river and were trying to get clear of the crossing before the fog cleared and the US Air Force came to call.

Before we reached the river we were shelled again, by UN artillery from a long way back. I had lagged behind my friends, so, when everyone dived into the ditches I kept walking down the road to catch up. A Chinese soldier, one of our guards, rushed up behind me and smashed me into the ditch with the butt of his Thompson. He did it for my own good, although I wasn't thinking so at the time and, as the huge shells were landing very close, he was putting himself at some risk. That was the only time I was struck like that during my captivity. I did get kicked later, but not hard.

The heavy artillery shells landing at that time were the only ones I was aware of during the battle. Reports read since claim the US liaison officer with the Glosters, who would have been able to bring in heavy artillery support from the big American 155 guns, had been withdrawn a few days before the battle. So the Glosters had only the lighter guns of 45 Field Regiment Royal Artillery in support. Nevertheless, our artillery worked miracles with their support fire until they were forced out of range by the Chinese advance.

We reached the river and prepared to wade across. I took off nothing, as the problem of undressing and dressing was going to take too long. As we waded through the water, which in places was nearly waist deep, we scooped water into our mouths. I had a raging thirst, we had had no water since capture.

The banks of the river were littered with Chinese dead, most of them being downstream – there were some jammed against rocks in the middle, others lying in shallow water and on mud banks. It may have been the crossing place where about fifteen men of the Glosters had lain in ambush on the night of 22 April and caught hundreds of Chinese troops helpless in knee- and waist-deep water. The river was about one hundred yards across at the crossing. I heard the Chinese had made mass attacks across the river against the Glosters' machine-guns, only succeeding when our men ran out of ammunition.

After crossing the river we were rushed away into the hills before the fog lifted. There are memories of woods, with many deep bunkers into which we were jammed like sardines during daylight hours, when the US Air Force roamed free, blasting anything and everything.

The F80 Shooting Stars, ground-attack fighters of the US Air Force, to which we had recently looked for possible salvation, were now the terror of our lives. On a few occasions I thought I, personally, was their target. Having watched them circle our area we then watched with horror as they climbed, then flipped over and came straight toward us.

Luckily they had other targets in their sights, although on one occasion the other target was so close they released their napalm bombs before reaching us and I watched mesmerised as the horror weapons rushed overhead and disappeared from view to explode less than one hundred yards away. Unlike the tearing, roaring rush and crash of rockets, or the crackle and crash of cannon fire, any noise made by the napalm bombs falling was drowned by the sound of the aircraft engines. Strangely, the lack of sound seemed to add to the terror of the weapon.

We saw the horrific effects of napalm. Men with their clothes turned to ashes, hands like burnt sticks, ears, noses and lips gone. Eyes just wet reddened ash – and still alive – lying where they fell, waiting to die, as there was no help possible.

Every day more prisoners joined the column until it was about two hundred and fifty to three hundred men. The guards, of course, were increased in proportion. The stories of other prisoners who joined us confirmed our worst fears. The Glosters had been destroyed. It seemed all had been captured or killed. Prisoners from other units, notably the Royal Northumberland Fusiliers and Royal Ulster Rifles, were fewer and their experiences gave hope that the bulk of their battalions had escaped the human flood.

I, who had removed first my badge, then my beret, in the interests of self-preservation when picked on by a machine-gun maniac, was duly impressed and somewhat humbled by the Royal Northumberland Fusiliers who joined us. They were wearing in their headgear the bright red and white 'hackle' of their regiment, which sticks up proudly three or four inches above their beret and can be seen for bloody miles. I heard one of them saying it would take more than a few Chinese to make them take out their hackles on St George's Day!

Gradually we built up a rough idea of the catastrophe that had befallen

29 Brigade and the Glosters in particular. The stories were many and varied from the comical to the horrific. I noticed that any amusing incidents were recounted with relish; the horror stories were not really mentioned, except in the context of some larger incident and there were stories of heroic acts – always by someone other than the storyteller.

One good story going the rounds at that time came from the RNF. They told how some of them had been curled up with laughter at one stage of a battle as they watched their cook chasing a Chinese tommy-gunner round and round the cook wagon, waving a meat cleaver, as the unfortunate man tried to clear a stoppage on his weapon. If the final outcome of that situation was mentioned, I cannot remember it.

Some of the Glosters, whose position overlooked a few abandoned and disabled vehicles, told how a great cheer went up when the NAAFI wagon received a direct hit, blowing it to bits and scattering money in the wind like confetti. NAAFI was not too popular with the front-line troops on account of the extortionate prices they charged for bringing their goods into forward positions.

One man in the Glosters had been watched in amazement by several men in another platoon as he dashed through very heavy automatic fire at close range to collect ammunition, then ran back into the position with a heavy ammunition box in each hand to resupply his platoon which had been cut off and partly overrun.

Another man was seen to play dead, in a position which was partly overrun, until a tommy-gunner jumped into his trench, whereupon he disabled the enemy and used the enemy weapon and ammunition to carry on the war.

Almost everyone seemed to have run out of ammunition before capture. I seemed to have plenty of ammunition, it was the luck which ran out in my case.

Several men of the Glosters told me the same story, at different times. They had been taken prisoner, and were sitting at the roadside after all resistance by the battalion had been crushed. An obviously senior Chinese officer stepped down from a vehicle near by, surveyed the wreckage and havoc all around, littered with many Chinese dead, then walked over to the prisoners and said in perfect English: 'You are British.' It wasn't a question, more an accusation. He gazed around at the local scenery again, then said, 'Twelve thousand miles from home – and you fight like this. God help anyone who lands in England.' With that he jumped back into the truck and departed, having paid 29 Brigade the highest compliment it could ever get from anywhere.

As I was not there, I cannot vouch for the truth of the story, but I heard it several different times, from different men who were there, and the story was almost word for word on each occasion. So I believe it.

This was the first indication I had that maybe we hadn't made such a mess of it after all. If the Chinese thought we had been a nuisance to their advance perhaps we hadn't been such a dismal failure. Nevertheless, I could not imagine the British Army, with its proud traditions, looking upon us very kindly.

Very few, if any of us, had any idea of the strength of communist forces which had been thrown across the Imjin River to destroy us. It had obviously been a big attack involving thousands of enemy troops but further than that the average trog, or even the average officer, knew very little.

We referred to various places in the battle as 'our first position' or 'our second position' and the place which, a few weeks after the battle, was known to all the Free World as 'Gloster Hill', was only known to us for the next two years as the 'Last Hill'.

Only fifteen men of 'B' Company had survived our second position to fight on the Last Hill, the rest had all been killed or captured, except three. One was severely wounded but taken care of by Korean villagers until UN forces returned about one month later. The villagers had continued to look after the man even after one of them had been executed by the enemy for taking food to him. Which says a lot for the conduct of British troops in foreign lands! The other two 'B' Company men managed to get through the enemy lines a day or two after leaving our second position.

We knew nothing of what happened to the masses of enemy troops as they poured south past our positions. How the attack was ground to a standstill and finally destroyed by UN forces whose time to get into position had been gained by our seemingly wasted efforts.

Not for years after the event did we learn of the great battle twenty-five miles to the east of our positions, where the Commonwealth 27 Brigade – Australians, British, Canadians and New Zealanders – put up a terrific fight, during those very same days that our battle raged, helping the US 1st Marine Division to destroy part of the Chinese offensive. In fact, the Canadian battalion ('Princess Pat's') was surrounded by enemy forces and cut off for twenty-four hours, so it must have been a close thing for them too. Luckily for 27 Brigade, although they were heavily outnumbered, there was little more than one Chinese division hitting them, so their situation was not as hopeless as ours.

On our left – to the west of the Glosters – the ROK Army 1st Division had held firm under heavy attack. But as the enemy poured through our positions on 25 April they were forced to redeploy under threat of encirclement. To the east of 29 Brigade the US 3rd Division also had to hinge back its line for the same reason.

Had we known of these events, especially the action of the Commonwealth Brigade whose effort against overwhelming odds did much to smash the attack and stop the Chinese advance on their part of the front, our morale would have lifted considerably. As it was, all we had seen was seemingly endless masses of enemy forces which poured over us in apparently unstoppable numbers.

At trog level we knew nothing about having achieved the impossible. That (militarily) we were doomed from the outset of the battle and stood no chance at all.

We had done what we could and been beaten by the enemy. That fact was only too obvious. We were shabby, unshaven, unwashed. Most of us were wounded. We were cold, aching, sore and very hungry. Physically and mentally we had taken quite a hiding, our pride was severely dented, and yet the trogs' morale was comparatively good. Not high, but high enough. We could still laugh at our plight and make jokes about it. Spiritually, I realised, the British trog could never be beaten. As individuals, I think we all plumbed the depths of despair, but as a group we lifted the individual's spirits back to group level.

About seven or eight days after capture we received our first food. It was a great tub of white, stodgy rice. Two Chinese carried it on a long pole and placed it on the ground in the middle of the area we were resting in. All the men were starving and there was a mad rush when we realised it was food. A young officer tried to get them to form an orderly queue, and finished up sitting in the tub of rice, which must have burnt his tail a bit, as it was pretty hot.

There was no way the more severely wounded could join the scrum around the rice tub so we just sat and watched. George Baker saw the problem and dived into the scrum to bring us our share.

There was no discipline in that state of hunger, it was every man for himself. Many fingers were burnt in the hot rice – but at least it put something in our bellies. After that first meal there was some hard talking done. It was decided that in future we would organise food distribution so that everyone got his share.

Eating rice, sometimes sloppy rice, was not as easy as it sounds. Practically no one had any spoons or other eating utensils. We had no plates or mess-kit. So we had to do what we could with what we had. Like many of the others I used the inside pocket of my tunic for a plate. George Baker tore it out for me. For a long time I used my cap badge as a spoon. Have you ever seen a Glosters' cap badge? Well, it makes a bloody awful spoon. The only reason I had the cap badge was because I had taken it out when being shot at and put it in my pocket. My beret was long gone, left in my last firing position.

About five days after capture one of our medics had gone around all the wounded, changed their dressings and cleaned up the wounds as best he could. It was then that I found I had two bullet holes in my shoulder, about two and a half inches apart. There were no exit holes at the back, but my left side was still on fire and the medic guessed at least one of the slugs had glanced off bone and gone down my ribs. He reckoned at least two ribs were broken. The pain stopped him from probing too hard.

Breathing was still difficult but the medic reckoned my lungs were okay or I'd have been spitting blood at some time before now. The worst pain was in my left hand. It felt like it was permanently exploding from the inside. Every heartbeat was like a kick on the funny bone. Someone gave me an army glove to put on the hand. It was one of those with a felt back and leather palm. Anything touching the palm of my hand gave me a terrible pain, but once I had forced the glove on, it felt better and life became more bearable.

Almost every prisoner had some sort of wound. Many of us had collected a lot of shrapnel which, when it went septic, came up in boils which were painful – probably more painful than when the shrapnel went in. It was in my case, for sure, as I had not felt most of my shrapnel when it entered.

When I looked around at some of my fellow prisoners, my couple of nice clean bullet holes seemed nothing compared to their wounds. One had a lot of shrapnel in the top of his head. I can hear his screams now as his friends held him down and tried to give him some relief by pulling the worst pieces out with a pair of ordinary pliers. No anaesthetic of course.

Another young lad would give subdued screams and moans for the first two miles every night, until the stiffness went from his wounded leg. He had a large bullet hole through his thigh. Others had smashed hands, smashed wrists and gaping wounds were quite common. Several had bullets in their chests and shoulders. One, with several bullets in his chest, spat blood every night as he marched the first two hundred miles. Eventually, he was put on an ox cart, but died within sight of the Prison Camp.

The 29 Brigade prisoners were in two columns for the march north. Two or three times we passed each other and many friends were glad to see others they had thought dead. I was damned glad to see Vic grinning at me the first time we saw the other column; we had not seen each other since he returned to the position after helping carry the wounded tommy-gunner to the aid post. He had been moved to another section, where there had been casualties early on.

Just after seeing Vic and a few other friends I'd thought lost, I saw

Colonel Joe Carne. He was standing apart from his column, greeting all the men in my column as they passed. As I approached him I was very surprised when he said, 'Well done, Large. Chin up, we'll be okay soon.'

What a man, how could he remember the names and faces of nearly all his battalion, even to new arrivals like myself – who certainly had not looked like this the last time he had seen us. We now had about three weeks' growth of beard on our faces too. The Colonel was obviously very pleased to see so many of his men alive as, until then, I believe he thought the one column were the only survivors.

By day we hid from the UN jets. Once away from the front area we were not put into bunkers any more, but allowed to rest in sparse woods or scrub. Occasionally the jets would strike at targets close by, bringing the war a bit too close for comfort again.

At night we would cover about twenty miles at a fairly fast pace. Sometimes we marched north all night, sometimes east or west and once or twice even south. Roughly every hour, sometimes two hours, the column was stopped for a five- or ten-minute rest. We never knew how far we were going or what time we might pause for a rest, or reach our nightly destination.

We soon twigged on to the Chinese orders for rest or sleep-stop. Someone who accompanied the officer in charge of the column would shout an order from the front and this would be repeated by each guard in turn back to the tail end. As we always moved in single file the column was quite long.

A shout which sounded like 'Suselah-Ah' meant 'take five'. The one I always longed for was 'Sweejo-Ah'. That meant we had reached our destination for the night and would be herded into a small area, to be surrounded by guards and allowed to sleep. Many of us often fell asleep during the five-minute breaks. I remember once falling asleep on a pile of loose rocks, in pouring rain. To get us moving again the guards shouted all sorts of things. The only word I can remember sounded like 'Kukali'. Whether this was Chinese or a botched version of 'quickly' I've no idea.

The guards all had a very frustrating habit of yelling loudly and pointing vaguely with their chin, in oriental fashion, often leaving us guessing what was required – and the wrong guess could bring a torrent of full-volume screaming in Chinese, accompanied more often than not by a lot of very rude stabbing motions with the bayonet. They had obviously copied an old British Army saying, but, in this case, they thought everyone could understand *Chinese* if it was shouted loud enough!

Our morale on those night marches was, I believe, quite high considering our circumstances. There was a typical trogs' stubborn acceptance of everything and, although there were many of us quite severely wounded, no one – as far as I know – was more problem to his comrades than he could help.

Everyone helped everyone else. I know the human mind tends to blot out the bad experiences and remember the better ones, but through the haze of agony which dominates the memory of that grinding, stumbling trek northwards through the dark mountains and hills of Korea I remember the songs.

The Glosters had been stationed in the West Indies in the late 1940s and (strangely to my ears) there often came floating through the darkness the sound of several voices (some melodious, some otherwise) singing the calypso-style songs of the West Indies. Such songs as 'Stone Cold Dead in de Market', 'Brown Skin Girl' and 'The Blue Tail Fly'.

One song, probably not from Jamaica, I will always associate with memories of that march. I think its title may be 'Two Little Lambs'. One man, whose name I forget but whose voice will never be forgotten, would twist the words a little and sing out: 'We're a few little lambs, who have lost our way.' Then perhaps fifty or one hundred voices would join in the 'Baa! Baa! Baa!'

The sound of that song comes back to me from the steep rocky slope of a mountainside, from the wide expanse of a night sky full of moon and stars and from the roar of pouring rain on a twisting little track through jumbled rocks. Just when things seemed to be at their worst someone, usually the same man, would kick it off – and lift the dark smother of despair to get existence into perspective again.

Our Chinese escorts varied in their reactions to the songs. Some, usually the older, hardened veterans, would take it all with a grin. Others would rush at the column, screaming at us to shut up and making a lot of threatening motions with bayonets, rifle butts, boots, rocks – anything they could lay their hands on. Even there, on that long grinding hoof northward, one of my great hates reared its ugly head: bullshit. Not bullshit itself, but its consequences.

Before leaving UK the units of 29 Brigade had been in the usual position of having to do the average amount of bullshit which their situation demanded. To the uninitiated it may sound madness but almost every trog in the army in those days found himself compelled to bull his best boots to a glittering shine. Many tools were used to smooth out the natural wrinkles in the leather so as to produce an overall smooth surface on which to work for hours with cloth, spit and polish (literally). The tools to smooth out the leather were various, such as toothbrush handles, knives, knife handles and – most potent of all – red-hot pokers.

A few days before the Imjin battle some of the troops, who had worn out their boots, had changed to their 'best boots'. Having been captured, these men were marching in boots which, a few months previously, would have kept them out of trouble by shining brilliantly. But now, those same boots, in their bull-damaged state, cracked up under the strain. The leather split across the top of the foot and the edges of the material cut into men's feet as they marched. I have seen one or two men, whose feet were cut in this way, with blood oozing through the cracks in the leather. As if they didn't have enough trouble already. In the footwear department I was lucky. Having been captured wearing a good pair of FP boots (Finnish Pattern, winter-warfare boots) and two pairs of socks, I had few problems. One thing which really got me were the cuckoos. Yes, in April, cuckoos go to Korea too. Being a country bumpkin it sounded a bit too much like home, except they sounded like they were laughing at our stupid human misery. Another touch of home was the magpies. It amused me to see they had built their huge nests in the telephone wires because of the shortage of trees in some areas.

At night the bombers came. We had been told that the US Air Force had destroyed every bridge and stopped all motorised traffic between the Korean front line and the Chinese border.

On our first night after crossing the Imjin River we were surprised to see a vast convoy of trucks on a road, all with headlights blazing, coming south. Some of the men thought it had to be the Yanks: 'Nobody else has lights on like that in a war zone!' But the Chinese did. It was at first a great mystery to us how they could get away with it, when American bombers were liable to be along any minute, looking for just such a target.

When we occasionally marched along a road at night we found out how they did it. Every few hundred yards along the roads the Chinese or Koreans had a sentry post, and on every truck that moved there was a sentry with a rifle. When a sentry post at the roadside heard a bomber he would fire over the next sentry post, which would fire over the next, until the sound reached the convoy when, because of the noise of the trucks the sentries on top of the trucks would fire in the air and shout 'Fiji' (which I presume is Chinese for aircraft). Upon which the trucks would immediately switch off their lights and pull to the side of the road, or right off it if possible.

To counter this ploy, the Air Force dropped flares to light up the road every so often, in the hope of seeing even one truck which might indicate a convoy. We spent many a sweaty time, lying in the ditch at the side of the road, alongside a damn great truck, with a flare hanging right overhead and a slowly circling bomber trying to make up his mind. Once or twice they bombed us. Once or twice they machine-gunned us, but by great good fortune they never hit any of our men. (As far as I know.)

At that stage in the game our bomb happiness had long gone. At least, I know mine had. The rushing whine of falling bombs, the savage, tearing shriek of shrapnel, or the thunderous roar of big 50-calibre machine-guns from a couple of hundred feet above me turned my guts to jelly. I sometimes longed for the means to hit back, forgetting they were on my side.

The guts of the aircrews had to be admired in spite of our circumstances. They must have suffered some losses from crashing into hills and mountains, as well as from the volumes of machine-gun and cannon fire which they often met. To fly along a road at two or three hundred feet at night, strafing and bombing, must have been pretty risky. The aircraft were, I believe, B26 Bombers.

The Chinese were past masters in the art of camouflage. The roads were all dirt, no tarmac. In the dry weather they became a ribbon of nearly-white dust. The trucks churned it up in great choking clouds which settled everywhere. Especially on the trucks and their loads. So, from above, the trucks were the same colour as the road and, as long as they kept still, were very difficult to see.

The same idea worked in reverse for the bridges. Most of the bridges between the front and Red China were intact, just as their Japanese builders had left them. We know because we walked over them. The reason the Americans thought they had destroyed them became obvious. They were covered in pine branches, not like a roof, but laid on the deck of the bridge and lashed to the rails or parapets each side. From the air, the white ribbon of road came to an end at the bridge and there was a dark gap, matching the river, where the bridge had been.

A lot of the trucks being used by the Chinese were American, captured US Army GMCs. I saw one British truck with the 29 Brigade sign on it. The vast majority of the vehicles, however, were Russian and ran on diesel. The smell of diesel fumes was strange to me, and it seemed nearly always that mixed with the smell of diesel there was the smell of death. The stench of death was everywhere along those roads. For many years after, every time I smelt diesel fumes I was reminded, and waited for the stench of rotting flesh to creep through. Even now, many years later, I sometimes think of those roads when I smell diesel oil. Somewhere, halfway up North Korea, when we had covered about two hundred miles in our wanderings, we came to a place known to us all afterwards as the 'Halfway House'. By this time dysentery had begun to take its toll and the column of prisoners was weak and weary.

Ten days after capture, a Chinese medical team had come to us, cleaned up our wounds and put on clean dressings. In my case they inserted the point of a glass tube full of iodine in the worst bullet hole, squeezed the rubber ball on the other end and squirted a lot of the stuff into the wound. Some of it came out of the other bullet hole. Then they covered both bullet wounds with antiseptic powder and did a neat bandage job over the lot. Most of the pain in the shoulder eased considerably after that. A good thing too, as no medic looked at it again for two years. The two British Army field dressings had lasted me ten days and prevented any serious infection.

The Chinese were understandably short of medical supplies, considering the casualties on their side at that time. About fifty miles back from the front I saw Chinese wounded – among them two on makeshift crutches and one blind – making their way north on a track through mountains. They may not have come from the front and were more likely air-raid victims; nevertheless, it made one think.

Before we reached Halfway House, I caught dysentery – along with almost everyone else. Being one handed and now considerably weakened, getting the slacks up and down was some sort of harassment. The guards didn't take kindly to anyone who broke ranks on the march, even when the cause was obvious. By this time we had changed guards a couple of times and now many more had rifles and bayonets, instead of the previous Thompsons. I seemed to spend half of each night's march struggling with my slacks – with an angry Chinese and a flashing bayonet right where I didn't want them.

Two nights before we reached Halfway House, I was in a state of collapse, alternately sweating and shivering and feeling like death. George Baker came to the rescue, half carrying me for most of the night. How he did it I'll never know. Half the night I think I was unconscious and delirious. In the bits that weren't blank, George was there, cheerful and encouraging. Sometime that night I had collapsed during a rest stop. I was brought round by a Chinese officer (not one of the guards) kicking me in the ribs – the broken ones of course. He wasn't kicking very hard but the pain was horrific. After a few seconds I passed out again. The next thing I remember I'm 'walking' along the road, supported by George Baker again.

There was a very strong suspicion that if you couldn't keep up with the column, you were shot or bayoneted, so there was a great incentive to go a little further. That night death would have been welcome. The next day I improved and by night had improved beyond measure. Unbelievably, I felt so good, the next march was a piece of cake – I actually enjoyed it!

So, the Halfway House. There were so many sick and weak among us the column had to be rested. Both columns of prisoners were rested there. The sick were separated from the not-so sick. After two or three days, those considered fit enough were marched north, leaving the sick at Halfway House to follow on about three weeks later. The Chinese told us nothing. We never knew where we were going, how far, or anything. When the fitter ones marched off we didn't know where they were going, or if we would ever see them again.

The three weeks we rested at Halfway House we stuck to a daily routine. At night we slept in buildings, in my case a cattle shed with muck on the ground and very little else. At first signs of light in the morning we staggered off to a small wood about half a mile from the buildings. Buildings were unsafe during daylight because of air attacks. We were fed once every day, a small quantity of rice in the evening.

During our stay at Halfway House, we were given one sheet of flimsy paper each and told to write home. None of us believed this paltry bit of rubbish, like cheap toilet paper, was really going to go back to the UK. Some refused to write, saying it was obviously some sort of Commie trick. Others, including myself, wrote home.

I sat for a long time looking at that piece of paper, weighing up the pros and cons. If I wrote and told the truth it would be very unlikely to get out of North Korea. Also, if, by a miracle, the letter did get home, what good would it do? My family would be worried sick, to say the least. The choice was obvious. I wrote a letter saying I had been slightly wounded and taken prisoner. I was being well treated and not to worry.

The date then must have been late May. On my twenty-first birthday, 27 September 1951, the letter arrived home, along with another letter written later in the PoW camp. It was the first information the British Army had to say where I was. I had been reported 'Missing'; 'Killed in Action'; 'Missing, believed PoW'. So my folks had no idea whether I was dead or alive. The War Office (now Ministry of Defence) requested they send them the letters and that they certified the handwriting was mine.

Another thing happened at Halfway House. The Chinese suddenly descended upon us and took away every item of any possible value we had with us: watches, rings, wallets, pens, even pencils. They made a list for each man and promised it would all be returned 'in good time'. Some of the men saw what was happening, destroyed their watches and pushed them into the dirt and out of sight. I didn't get the chance or would have likely done the same.

The excuse for this caper was that some prisoners had been giving away valuables in exchange for tobacco from the civil population. This was very true, in fact, as we would do anything for a smoke. I saw a gold watch exchanged for some raw tobacco leaves. The Koreans grew a lot of their own tobacco, or at least I presume they did, as we could often get tobacco (*tambi*) leaves from them.

Paper was at a premium for making cigarettes and I have seen men taking leaves from the Bible to roll a smoke. The Padre went right off them when he found out.

One of the most expensive smokes I saw was rolled in a five-pound note. I rolled cigarettes in up to ten bob (fifty pence) notes, but drew the line at a quid. Money was no use to us any more. Only some of the American prisoners seemed to value money. We never used it after capture, as far as I know, just kept what we had and forgot about it.

Our money was not in local currency but in British Armed Forces Special Vouchers, which everyone called 'Baffs'. The values were the same as sterling but the only coins were plastic pennies, everything else was paper money. Starting at threepence there were notes for sixpence, one shilling, two shillings and sixpence (half crown), five shillings, ten shillings, one pound and five pounds. There may have been notes of higher value but I never saw any.

I think my pay was about six pounds per week in Korea. The average civilian wage in UK at that time was probably ten to fifteen pounds per week, maybe fifteen to twenty pounds per week. When you consider twenty cigarettes cost one shilling in those days, (five new pence), smoking even a threepenny note was not good value for money!

The surprising thing about our valuables was that we got them all back. We had been at the PoW camp a month or two, maybe three, when one day we were all called to the parade square. We were given the usual lecture about how well the Chinese People's Volunteers treated us, then they produced a pile of packing crates, opened them and inside were our valuables. Each set packed in paper and the crate packed with sawdust to prevent damage. Pickfords would have been proud of them! As far as I know, only one man complained he was 'robbed'. He had a wedding ring returned to him but claimed it wasn't his. I thought at the time he had possibly lost so much weight it didn't fit, or look right any more, but I never found out. The Chinese would have got a lot more value from the whole episode if they hadn't been so busy with the brainwashing at the time. Everything they did was viewed with the utmost suspicion.

Before leaving Halfway House, some American prisoners joined us. We already had an American pilot there who had been shot down and was badly burned when he baled out. Now we had about half a dozen more US personnel, another pilot, I think, and one man in particular we called

'Poncho'. Poncho was a tank driver and was very bitter about his unit leaving him to be captured after his tank hit a mine. He was Mexican-American, short, stocky and very powerful for his size.

Being tall, my obvious nickname in the forces was 'Lofty'. Poncho, however, always called me 'Lefty' as did many other Americans. Lefty was a common nickname in the US whereas few of them had heard of Lofty.

As is often the case with opposites, Poncho and I became good friends and remained so throughout our captivity, although we did not see each other often, and I have not heard anything of him since my release from the PoW camp. He was amazed to hear how things were in the UK. Like most Americans at that time he thought, for instance, that the police in the UK had very few cars, perhaps none at all, and when chasing the criminal element they would jump on the running-board of a passing car and shout 'Follow that car.' In fact, I once settled an argument between several Americans, half of whom were fully convinced that running-boards were a legal requirement on cars in UK for the police to use in emergencies.

Poncho was full of funny stories from 'back home', both army stories and hilarious Americanisms. He was very good at unloading dead time.

A couple of our boys tried to escape from Halfway House, but were caught and dragged back looking somewhat the worse for wear. One was tied so tightly he had the rope marks on his arms years later.

We spent almost all our time swapping yarns and delousing our clothes. Lice were our constant companions from the first week after capture. At some time during our 'holiday', the Chinese turned a couple of 'barbers' loose on us. They were Chinese and obviously obeyed orders to the letter. Having undoubtedly been ordered to 'shave the prisoners' faces' they did so, quite expertly, starting at our hair line they shaved every bit of our faces – removing our eyebrows and eyelashes on the way. Had they shaved our heads we would have made a good set of snooker balls, but they didn't touch our hair. No haircuts, just a complete face shave!

About three weeks after our arrival at Halfway House we were formed up in a line one night and marched off to the north. No warning was given, we never knew where we were going, or when. Soon after leaving we started marching by day, the risk of air attack became very much less as we moved further north. Somewhere along the way we had to wade a river and were amused to see a huge rusty Russian tank perched high on part of a bridge. The bridge had been destroyed at both ends. The one remaining part – with the tank perched upon it – was tilted at a steep angle like a seesaw.

On that last half of the march, mostly in daylight, we had the feeling we were being paraded for the benefit of the civilian population. Instead of avoiding towns and villages, by using faint tracks through the mountains as before, we now seemed to be forever in front of hostile crowds. Perhaps the change to daylight marching made it seem that way.

At one stopping place a medical corporal was kicked by one of the Chinese officers and so he made a threatening motion with a stick. He was immediately pounced upon by several guards, his arms lashed together behind his back, then, with his arms out straight behind him he had a rope tied round his wrists and was hoisted off the ground to dangle from the branch of a tree for two or three hours. We watched, helpless to do anything, as he went in and out of consciousness, the dysentery dripping from his feet, while the guards glared at us over their guns. This was a very isolated incident, and was most likely done to impress the rest of us with the benefits of good behaviour rather than to punish one man for his actions.

Eventually, we came to a large cultivated valley with a small town in the centre of it. We marched through the town to its northern outskirts and found ourselves in a suburb which had been turned into a prison camp. No barbed wire, just sentries around the perimeter. As we stopped to survey the scene a big red-haired American left a group of his fellow countrymen and rushed towards us shouting: 'The Brits have come, the Brits have come!' I think he thought he was being liberated. There was some yelling from one of the guards – which had no effect – then he was shot, falling dead in the gutter a few yards from us. A British voice from the nearby huts called quietly: 'Welcome home.'

4 PRISON CAMP

The PoW camp at Chongsung, in North Korea, was in the northern suburbs of the town. Between the camp and the main part of the town was the Chinese Headquarters, housing the guards, administration staff and all the hangers-on required to run a PoW camp. The main road to the north ran through the camp, which had no gates, barriers or fences of any kind. There were no fences or barbed wire anywhere around the camp, just a lot of guards, spaced about forty or fifty yards apart all around the perimeter and anywhere else we were not allowed to go. Manpower was cheaper and more effective than barbed wire in China. The guards stood between thirty and fifty yards outside our perimeter (usually the outer edge of a road) and shouted a warning if we ventured near it. After the shouted warning – in Chinese of course – if you did not retire quickly enough two or three weapons were aimed at you to make the point clear. I never heard of anyone pushing their luck any further than that. A river flowed north along the west side of the camp. To the east the ground sloped upwards fairly steeply to hills, backed by distant mountains. The Land of the Morning Calm can be very beautiful in that area, and it was so at Chongsung.

American prisoners were housed in the northern part of the camp, British prisoners in the part nearest the Chinese garrison. Houses in Korea were of simple construction, but very practical for the climate. All walls were mud and wattle, interwoven branches with mud plastered on and through to form a smooth finish both sides. Usually the inner walls were papered thickly with newspapers or picture magazines. The roofs had a wide overhang to keep rain from the walls. Most roofs were shingle (wooden planks used like tiles). All the houses were constructed with a strong wooden frame of large beams and posts and all were bungalows.

A cooking fire was below ground level, in a pit about one metre deep at one end of each building. The chimney of the cooking fire was a network of tunnels under the length of the floor to a chimney stack at the far end. By this means the floor of the house was kept warm with very little fuel and so, as the rocks used in the construction of the floor retained the heat for a long time, it warmed the whole house. The floors were smooth mud on top of the rocks and this distributed the heat more or less evenly. Each house usually had three or four rooms. Access from one room to another was usually via an outside verandah along the length of the building. Most verandahs had raised wooden floors.

There were few, if any, windows in the houses. I don't remember seeing any in my part of the camp. The doors to each room were lattice, usually from top to bottom, and sheets of thin rice paper were glued to the inside. This allowed enough light to enter the small rooms to let you read except, perhaps, in the furthest corners. I saw no glass windows anywhere at Chongsung.

Most of the chimneys, being a long way from the fire, were constructed by nailing together four long planks of wood to form a square-sided tube which leaned away from the building and was supported near roof level by metal or wooden struts between chimney and building. I think all the buildings had the fireplace at the north end and the chimney at the south end, presumably with regard to the prevailing wind in the area.

The PoW camp had the narrow streets and alleys of its original intended use, as a part of the town, with open areas here and there. Along the streets were monsoon ditches, lined with rocks, to drain the frequent heavy rains. The ditches were usually one or two feet deep and one or two feet wide, but sometimes deeper and wider.

As soon as we arrived we were organised into squads of ten men and allotted our rooms. Everything was 'communistically' fair. Ten men – one room. In my squad of ten there were at least three of us over six feet. Our room was a small one, six feet six inches by six feet. Our lice were all related and likely suffered from inter-breeding. To make things worse, I could only sleep on my right side as my left arm and ribs were still a mass of pain. As there was no room to sleep on our backs and if one turned, all turned, there was an obvious problem. It says a lot for the men in that room that we had few harsh words in several months of those conditions. Another squad nearby had a room almost twelve-feet square and there wasn't a six-footer among them. We were not allowed to change rooms and had to be in our own room to be counted at any time of the night.

The British contingent was organised into two companies, each of three platoons, each platoon had up to ten squads with up to ten men in each squad. Each company had its own Chinese company commander and staff who worked in the company office, which was shared by the Chinese platoon leaders and 'instructors'. The Chinese platoon leaders, one for each platoon, could not speak English, but gave the orders. The Chinese instructor could speak English and interpreted for the platoon leader. He was really a fairly well-trained political commissar whose

English was often not that hot. In many cases – thank goodness!

Each company had its own cookhouse with one Chinese cook helped by two or three prisoners. Daily routine in the camp started early. A Chinese platoon leader would emerge from the office on the dot of 6 a.m. and blow his whistle loud and long. Then he would walk around the huts blowing the whistle from time to time. As soon as the whistle sounded we were expected to move quickly from our huts to the company parade ground to be counted.

After this initial roll-call the Chinese company commander would usually give a short speech, interpreted by one of the instructors, ticking us off for some minor laziness, or praising us for some achievement in the cleaning or wood-collecting jobs. Then we would be told what our next trick was. It was usually to march up to the main parade ground to join all the other prisoners, British, American and any other nationalities, for a long and boring political lecture.

The lectures were never a known quantity. They could last an hour or all day. We never knew when we would return to our huts. If we did get a short lecture it was pretty sure we would be instigated to do political study in our rooms. The Chinese would roam from room to room trying to get a political argument going, listening outside the rooms to check if we were discussing the correct subjects and generally making a nuisance of themselves when the last thing we wanted to do was talk politics. Sometimes the whistles would sound again and back we would go to the main square for another few hours of political patter. Those whistles ruled our lives. I still hate hearing whistles.

The officers were put in a separate camp on the other side of the town, along with American officers. After a week or two in the camp, all ranks above corporal were moved to another camp near the officers' camp. I believe there were roughly six hundred British prisoners counting all ranks, officers, sergeants and other ranks at Chongsung.

At a guess, I would think the American prisoners numbered about three thousand. Some of them had been captured in 1950 and, by the time we arrived, a lot of them were in a pretty bad way. There were only two British prisoners in the camp before the arrival of those captured in April 1951. They were two men of the RUR who were caught in an earlier battle. One had been wounded before capture and by the time we found out he was there – two or three weeks after our last group arrived – he was at death's door. Some of our people approached the Chinese 'instructors' and the man was moved in with the British contingent, where his friends from the RUR took him on and worked hard to achieve a good recovery. We had a very small sugar ration at the time and every British prisoner must have given a small proportion of his ration at some time or another to help someone who was 'going down the pan' as we called it. Dying.

When he first came to us, the man from RUR was deaf, blind and totally crippled. Within a month, thanks mostly to other RUR prisoners, he could see and hear. In two months he could walk again. All through our captivity the British prisoners helped each other. The fit ones helped those less fortunate. No man felt alone for long. It just happened, without any apparent organisation.

The American prisoners had been captured before us. The winter in captivity had taken its toll, not only in lives but on the morale of the survivors. They were dying at an alarming rate. I certainly did not spend my time watching burial parties, and I was not particularly interested in the problems of the Americans, but one morning I did notice twenty-five bodies being carried up 'Boot Hill' from the American camp. When I mentioned what I had seen, no one seemed to think it out of the ordinary. There were burial parties on the hill every day, but I never did count again.

Some of our friends who had gone to 'D' Company were missing, but we had high hopes they had escaped to the UN lines after the battle. We knew, from some of the 'D' Company men who were captured, that some of their company had managed to break out, although, when last seen by at least one of the prisoners, they were being shot at by American tanks. Some of the 'D' Company men who were captured had turned to take on the pursuing Chinese rather than fall under the guns of their allies. Years later I learned that thirty-five men escaped capture or death. None of my friends was among them.

Three or four weeks after I arrived at the prison camp, Vic went down with malaria, which he had contracted in India while with the 1st Wiltshires. It went the usual course of up and down, with the downs becoming worse until Vic was eventually carted off to the prison camp hospital.

The hospital was an old Japanese temple on a hill overlooking the town. In 1951 it was not really a hospital, just a death house. No doctor, no medicine, no hope. There was straw on the floor to sleep on, which was the only thing better than the rest of the camp.

A week or two after Vic went to the hospital we were told he had died of malaria. About a month later, two Americans came looking for Vic. When told he had died of malaria, they said he certainly had not, they said they had been in the hospital with Vic and he had recovered from malaria after about ten days, had said he was returning to the camp, picked up his kit and walked out. Vic was never seen again. He had just disappeared. There was no mileage for the Americans telling us he had walked out if he hadn't, so we had to believe them. The Chinese we questioned knew nothing except Vic had died of malaria when they told us. They said they had received a report from the hospital.

Brits did not die easily. As far as I know only three British prisoners died in that PoW camp – though there may have been one or two others. Not counting the man who died of wounds just before we reached the camp, I can only remember three. There was one, I didn't know his name, or what he died of. There was Vic. Then there was 'Old Pop', who was fortyish, older than most of us. He had been wounded, he had malaria, tuberculosis, beriberi, dysentery and, of course, malnutrition. Old Pop hung on with that lot for weeks. It became almost the first question of the day: 'How's Old Pop?'.

Eventually he died. It had to happen, the odds were beyond even Old Pop's fighting spirit – but it said a lot for those around him who had done their damnedest to keep him alive, never giving up hope.

Just across the street the American prisoners, by contrast, had what we called the 'three whole days and no candy' disease. They had, as mentioned earlier, been prisoners longer than us, had suffered the rigours of a North Korean winter in those conditions. But we could not understand how young men, many with the badge of the US Marine Corps on their jackets, could just sit down and die. No wounds, no specific illness in many cases, but no will to live either. On the other hand there were many Americans who had good reason to die and refused point-blank to do so. They, like us, struggled on with their wounds, beriberi, dysentery and malnutrition – the standard full house in that place – and cursed their fellow men who let the side down.

We know that, in some cases, if an American looked like he was dying his room-mates would be playing cards for his boots, his wedding ring, his wallet – anything. In many cases, after he died, he would be propped up so the rest could draw rations for him.

There was one huge black guy who carried a heavy bag around with him all the time. Other Americans told us it was full of gold teeth. When I saw the bag which was soft cloth, or canvas, with a string around it to keep it closed, the round bulge in the bottom was the size of a baseball. I never saw inside the bag so I'm not sure what it contained, but – knowing the conditions in the camp at that time – I'm pretty sure it was gold teeth.

It is likely no one will ever know how many died at Chongsung. Once, when the subject came up, I remember someone saying it was over eight hundred. I think that is a very conservative estimate. On the one day I watched the burial party, they had twenty-five bodies. Perhaps it was an exceptionally bad day and at the worst time, I don't know. But the burial parties went up that hill every day for three or four months after we arrived at the camp. After that it tailed off until there were very few burials in the last few months of my captivity.

Soon after we arrived at the camp we were lectured on the Chinese People's Volunteers and their 'Lenient Policy'. We were told none of the Chinese in Korea were regular soldiers, they were all civilian volunteers; and because we were basically working-class people, not responsible for the war in Korea – having been duped by the Capitalist Imperialist Warmongers of Wall Street – we were eligible for the lenient policy as applied to prisoners-of-war. That was why our lives had been spared and we were being so well treated. We had not really been captured, but liberated from our Capitalist masters by the Glorious Chinese People's Volunteers – who were even now marching forward to victory in the south. (I remember thinking it sure enough looked like it when I left the front!)

Having put us all in the picture, one man was appointed squad leader of each group. He was appointed by the Chinese irrespective of British Army rank, age or anything we could pin down. The squad leader was responsible for collecting rations and organising the squad to clean its room and immediate area. He was also responsible for the good behaviour of squad members and the squad as a whole. Not an enviable position.

The next trick had us rolling in the aisles. Each squad was issued with a question paper and ten blank sheets, one per man. We had to study the questions then write our answers on the blank sheet. It was the questions and possible answers which raised a laugh. One question was: 'Why should Rockefeller have all those millions of dollars?' The answers were very varied such as 'Why not?' or 'He worked hard for it,' or 'Somebody had to pay the workers.' Near the bottom of the page: 'Who started the Korean War?' Answers: 'North Korea', 'Commies', and one I saw even said 'You bastards'.

They had a day or so to sort that lot while we grinned and swapped jokes about the answers. Then the laughing stopped. The lot of us were ordered on parade and the gist of the lecture which followed was 'Running Dogs of the Imperialist Warmongers of Wall Street are not eligible for the lenient policy and will be moved to a camp where the lenient policy has never been heard of. In future all men will put their real names on their answer sheets, the squad leader will be held responsible for every sheet being correctly answered and returned to the office. Squad leaders who fail in their duty will be punished.' Many answer sheets, except the funnier ones from the first batch, had been used to roll cigarettes, or in the on-going war on dysentery. Most of those returned to the Chinese bore those famous names like 'Mickey Mouse', 'Tom Mix', 'Kilroy' and 'Aunt Sophie'.

Now the pressure was on. There was no way we could risk getting our own people punished. The hate of all things political started at that point. The British Army is a non-political organisation. Loyalty is to the Crown, representing the Country, irrespective of political parties. Hardly any of us knew anything of politics in those days – before the present mass media situation proved there's little to pick between the lot of them, except there are the extremists and the rest of us. Once the political nuts had us over a barrel they never let up. They appointed a further 'hostage' in each squad. He was called the 'monitor'. It was his job to collect, distribute, read and discuss all the political documents issued and return them safely to the office.

We did the absolute minimum to get by without causing reprisals. The 'instructors' would go from room to room and try to discuss politics. Some of us argued with them, not realising this was just what they wanted. On the political scene they could make black look white – and prove it!

I argued and continually had any point turned against me, until I went right down to basics, where I had firmer ground to stand. Then they were stumped, got angry and I had to back off once or twice or risk trouble. The first argument where I nearly went too far was with the 'Hedgehog' (because of his haircut). He had proved black was white again and I said, 'You are a political instructor.' He agreed. I said, 'You can prove to me – politically – that black is white. Okay, I'm a soldier, you take a rifle and give me one. We'll walk up yon hill from different sides and see which bastard walks down again.' I didn't think it was that bad but the Hedgehog nearly had a fit. For a moment I thought it was jail for me. But he calmed down and stamped off, still in a rage.

They didn't like it when we mentioned how the Commies in Russia had helped Hitler smash Poland. It was a fact they couldn't dodge. I also got close to getting into hot water when I likened the Chinese operation in Tibet to Hitler's annexation of Austria.

But the one which really had them running for their red books and never did get answered properly, yet didn't raise any tempers, was the simplest. I told them I had lived under real communism – in a barrack-room – where everyone was on the exact same pay, had exactly the same kit, treatment and facilities. After a year or so, some had nothing, some had a little money saved, one or two had bought bicycles. Why should those who spent all their pay on beer and cigarettes be jealous of those who were now better off? Should they now share the wealth of the barrack-room between all? Communism in its true form cannot work because all people are different.

They said this was simplifying the question out of all proportion but could not come back with a real answer.

The Chinese were completely convinced they were right. That South Korea had started the war. That all the people living under the 'Capitalist Yoke' were being exploited every which way and were itching to be educated and liberated. They tried very hard to educate us, but failed miserably. The more we learned of communism, the better we could shoot holes in it. Eventually, after about eighteen months, they gave up. There were still lectures and propaganda talks, but we sensed their hearts weren't in it any more.

One thing I could always understand about the Chinese: they were immensely proud of their country, as any man should be. They said the name 'China' almost with reverence. They were brave men in Korea, as we knew only too well. I had no hatred for the Chinese, they were honourable men who did what they had to. Communism was like a disease in them; it did a lot of good in China, only because things were so bad before. But the nature of the Chinese people is too strong to be held in rigid shafts for ever. When they step from the shafts and are given their head to go, as go they must, China will be truly great, in every way.

Political lectures became a daily routine. Usually the whole camp, British and Americans, was herded on to a great parade ground, told to sit down in lines, then harangued, first in Chinese then in English.

After a few weeks someone had the idea of stools, so we were all issued with a stool to sit on. This consisted of a short log of about eighteen inches, with a twelve-inch plank nailed to one end. These stools made it more difficult to sleep during the lectures, easier for the roaming instructors and platoon leaders to keep an eye on our attentiveness. Like most of the men, I finished up with an amazing ability to appear really into a lecture, but always have my mind elsewhere – or switched off altogether. I think some of us could be effectively asleep with our eyes open.

This has proved a problem ever since. As soon as someone starts a lecture, I start to switch off. Unless the lecturer is very good and makes it very interesting, there is no way I have a clue what it was all about at the finish.

Some of the lectures went on for hours, starting at 7 a.m. and not finishing until sometime in the afternoon. I have even seen a couple of Americans, who had presumably marched to the parade ground, manage to die during a lecture. (Obviously bored to death.)

At some stage in our march north – I think it was just before Halfway House – I found an old rusty cigarette tin. It was a 50 Woodbine tin, round, about three inches deep and two and a half inches in diameter.

After cleaning it with sand and water, it became my food and water can. I could collect my food in it and found I could balance it on my injured hand, which protruded from the sling, to eat in relative comfort. Thanks to the glove I had no trouble now with hot rice.

When we first arrived at the PoW camp there were a number of Puerto Rican troops living with us and getting rations from the same cookhouse. No matter how I complained I got one fill of stodgy rice in my can, while the little Puerto Ricans were filling their American steel helmets. They must have been getting about twenty times more food than we were, but Communist justice was one container each. I longed to find a five-gallon drum.

After we reached the prison camp, rice became a luxury we didn't see very often. The staple diet was sorghum, sometimes millet. Sorghum has round grains which swell when cooked, as does rice, but it tastes foul and turns pink with blue and mauve patches when cooked. Millet was some sort of birdseed. That too tasted foul, even worse than sorghum. Our diet was boosted by a ration of sugar. This started when we first reached the camp and it was issued every ten days. The sugar ration for ten days was, at first, five level dessert spoons. This was increased to ten level dessert spoons, over two years.

After a month or so at the camp there was great news. We were to have vegetables to go with our sorghum or millet. In the event it was turnip tops, or something similar. We called it 'greens'. The greens were boiled with a lot of water in a boiler tub and, at first, the supplies of them were pretty few and far between. But the water had a certain taste and helped with the sorghum. By the time the greens came along, some of us had beriberi. The first signs are puffy feet and ankles. If nothing is done about the diet, the puffy condition moves quickly up the legs. After a week or so the swelling becomes painful, searing hot pains shoot up and down the legs, especially the lower parts. No one seemed to know the cure. The Chinese were sympathetic, but seemed to have no idea what caused beriberi or how to cure it. I'm quite sure they would have solved the problem if they could.

If nothing is done to arrest the condition then it ultimately leads to heart failure. We tried exercising, running and walking but this only caused a lot of pain.

In my squad we had a great idea. I was the only one with beriberi at the time, so, full soldier's logic was brought to bear on the problem. Everyone had ideas which I tried one after the other – unsuccessfully. Except one idea that I drew the line at. We had all decided the swelling must be fluid, so one idea that was that I could cut a couple of holes in my feet and just drain it out. I said that was definitely the last resort stuff – and anyone could try it when I got too swollen to move, but to make damn sure I couldn't move first.

Kicking that idea around, someone had a brainwave. The obvious answer – why hadn't we thought of it before – was to sleep with my feet up, drain it out of my legs and piss it out in the morning, simple.

So, that night my feet were propped up and I spent a more uncomfortable night than usual, not having room to sleep on my back of course. In the morning, shouts of victory when my propped up feet were seen to be greatly reduced. Then they saw my head. I couldn't see out of the swelling for about an hour after sitting up. The things I did for medical science.

Comforting words like, 'Jesus Christ, look at his head' and 'Blow your nose – it'll come out your ears.' And one who really tried: 'It's mozzie bites on his eyelids – they were murder in here last night,' did nothing to quell the panic I tried not to show. Afterwards I saw the funny side, but, at the time, I wondered what damage could be done in my head by this creeping, horrible fluid.

Our salvation came from a little bloke in the Medical Corps called Janman. He had read about beriberi somewhere and felt that he might be able to remember the cure. He racked his brain for weeks, from when the first cases were seen, among the Americans. One day Janman cracked it. The word came around like wildfire. 'Eat greens, eat anything green – grass, nettles, any damn thing.' Everyone with beriberi immediately went on extra rations of greens from the cookhouse. I ate grass, nettles, leaves and weeds of all sorts. We went through the camp like locusts. Beriberi was gone within a couple of weeks. By the time young Janman got it right I had fluid up to above my waist and was in some considerable pain, mostly in my legs. Walking was becoming difficult but I was afraid to lie flat – knowing it could creep up in the night.

Another problem which affected quite a lot of people after about eighteen months in captivity was night blindness. Although they seemed to have their normal eyesight during daylight, the moment the sun set they were almost totally blind and had to be led around by those of us who could see. No doubt it was another vitamin-deficiency problem, which was solved soon after their release.

I did notice that the people who suffered the problems of beriberi were nearly all men who had not been in the front line for long before capture. People like myself (perhaps a bad example) who had been used to a lot of rich food – lots of eggs (at least five or six per day) bacon and bananas, in Hong Kong, followed by terrific Aussie rations in Japan – had gone from that excess of nutrition to almost nil in one sudden step. Whereas men who had been in the field for months had been on much less nutritious field rations and the change of diet was not so extreme.

There was a lot of beriberi among the Americans, due perhaps to their rather lavish field rations. Our name for the 'disease' which killed many of them – the 'three whole days and no candy' disease – was, quite possibly, a good diagnosis, except it should have been 'Three whole months . . !'

Almost everyone had dysentery. Men had to rush out in the night several times, heading for the latrines. The latrines consisted of a big trench with poles around the edges. We sat on the poles 'Gloster fashion' – back to back. One man who never made it to the latrines became famous as the 'Phantom Arsehole'. Almost every morning we would find where he had struck, usually on the small parade ground between the huts and the latrines, sometimes on someone's verandah or in the street.

All the budding Sherlock Holmeses got to work on the problem, but the Phantom Arsehole eluded all for many weeks. Then we had our first mail from home, and the next morning when we turned out, we found the Phantom Arsehole had not only struck on the parade ground but, having put an envelope to good use, had, in the blackness of the night, left his name and address in the middle of the strike. So ended a sad mystery, which had whiled away many an empty hour and caused more laughs than it probably deserved.

I had at least one letter in that first batch of mail. Unknown to me at the time, Ann, my future wife, was writing five letters every week from the time I was first reported missing. Using all sorts of addresses which were sent to relatives of those missing, and some from commonsense, she bombarded North Korea with a deluge of mail – and some got through. During my captivity I received about one hundred letters. The vast majority were from Ann, but I also received a few from my mother and one or two from other relatives and friends. So about four hundred of Ann's letters went missing. She wrote to some strange addresses before a proper address was supplied by the Chinese.

On several envelopes my address was: my number, rank and name, 1st Glosters, PoW camp, North Korea, c/o Peking. Two or three letters were c/o Budapest. Ann used to put in the odd snippet of sports news – to fill up a page, no doubt. Although there were doubtless sporting events of great significance taking place in the outside world, the sort of thing I got was more likely a few results from Third Division South, a new name on the bike racing scene or when so-and-so beat so-and-so at cricket in the local village league. Nevertheless, these snippets were usually home news to someone in the camp.

Compared to most of my fellow prisoners I had a lot of mail. Why so much mail was stopped from reaching us I have no idea.

To have the best chance of getting my letters out of Korea I always gave the impression we were being well treated and there was no problem except waiting for an end to the war. One of my theories on getting letters out was: if the outside world knew we were there, then the communists would have to account for us one day – if communism was to be given any credibility at all in the free world.

Letters from home were, for many reasons, one of the greatest morale boosters we had. Just an envelope with our address in familiar handwriting would do wonders, even had there been no letter inside. In autumn of 1951 we started collecting timber from the mountains. The timber was for our own use, for cooking and heating during the coming winter. The wood details, as they were called, were an immediate success for many reasons. They got us out of the camp for much-needed exercise and we found edible fruit growing wild in the mountains and supplemented our rations.

The exercise was great for those planning to escape and allowed chances for local area recce for quick routes into the mountains.

At first light, a company of camp guards would double march off into the hills to surround the area of the wood detail. About an hour later we would march out into the appointed area, sometimes where trees had been felled and cut into lengths for carrying, sometimes into an area where we had to search for dead wood. I found my ability to carry anything heavy was severely restricted by my injuries, and was afraid of being taken off the wood details. But the Chinese showed a bit of commonsense and I was allowed to carry smaller loads.

I was still suffering considerable pain, especially in the left hand. Little did I think that some welcome pain relief was at hand – due to the wood details and Poncho. The Brits and Americans had now been separated properly, with guards posted at intervals between us. But, with a bit of careful watching, we were able to nip across the street without being noticed, especially if a diversion could be fixed to keep the guards busy. One day, it must have been sometime in late October or early November 1951, I heard Poncho had been looking for me in our part of camp, and sure enough a day or two later he was back – looking secretive, but like the cat who stole the cream.

He got me out of sight of any Chinese and out of earshot of anyone else then he said, 'Lefty, you still got the arm problem?' I nodded. 'Okay, now listen, your problems are over – I got the weed'. He looked at me like he'd just explained everything. Seeing the blank looks he got on with it. 'The weed, you dummy! It grows in the goddamn mountains – and I found it – it's the best goddamn pain-killer ever sent down the line!' I had a mental picture of Poncho rubbing my arm with some sort of herbal concoction. I obviously looked doubtful. 'You can't get hooked on it – if that's your problem. And it's the real McCoy, I've tried it myself. It's just like back home, there's no big hook, it's okay. You can take it straight or mix it with tobacco – a couple of pulls and powee! No pain!'

The beginning of understanding twitched in my thick skull, but I didn't really understand all this about 'hooks' and 'straights' – sounded like boxing talk. Poncho's patter always did take some to keep up with, but now he was excited. Full realisation only came when he pulled out this bundle wrapped in paper, about the size of my fist, and showed me what I had to do with the brown leaves inside. Poncho racked his brains for my sort of language then said slowly, 'It's marijuana, the weed. You smoke it. It's non-addictive, I promise you.'

My complete ignorance of the whole drug scene was such that if Poncho had just told me what it was and not told me how to use it, I would have very likely tried to make tea with those leaves. I don't think I would have trusted anyone else in that camp to tell me about using the weed. Certainly not my own people, who knew no more about it than I did. But Poncho I could trust absolutely in that department. Knowing something of his background, his knowledge of such things could be taken for granted. He looked the part anyway!

After Poncho went back to his camp, I found a quiet spot and remembering his instructions, built a smoke just of weed. Then I made one of my ordinary cigarettes and went to the cookhouse for a light from the fire. I lit the ordinary tobacco cigarette then, finding the squad-room empty, I sat down and lit the straight weed. At the first lungful of smoke from the weed my arm blew up. The pain was terrible for about four or five seconds, while I held my breath in panic. Then it went. I thought, well, shit or bust, nothing ventured nothing gained, and took another pull. This time, no increase in pain and within a few seconds all the pain had left my arm for the first time since being wounded. The relief was really fantastic. Unbelievable.

Remembering Poncho's instructions, I extinguished the weed and just sat there wallowing in no pain. I felt a bit stupid and couldn't stop grinning. I had plenty to grin about right then, apart from the effects of the weed; the pain was gone. I could feel like I used to feel.

The pain I had suffered during those first six months of captivity cannot be described or put into words. Some days it was worse than others, but it was there to be lived with all the time. I tried hard to shut it out of my mind and after several years have had some considerable success. Practice makes perfect, but in this case I now know there is not enough time for practice to make perfect in my lifetime.

Anyone who has had the usual problems of rolling cigarettes with paper and loose tobacco may pause to wonder how I managed with only one usable hand, coarse newspaper and very rough tobacco. Ability and expertise are often the children of necessity. I could roll a cigarette with one hand as fast as most people could with two hands. Not long before Poncho brought his great gift of pain relief, I had managed to move the little finger of my left hand for the first time.

Two of my friends in that place, both of whom had been PT instructors with the Glosters, had given advice and tried many ideas to try to ease the pain and get the arm to move. The biggest problem throughout was the pain. Any attempt to move the joints in the arm or shoulder produced screaming agony in the shoulder and palm of the hand.

I had discarded the sling soon after arriving at the camp. By then the arm was set in the position it had been held by the sling. By late autumn 1951, my arm and hand had lost almost all flesh. The elbow, wrist and finger joints were like knots in a rope. The fingernails of that hand never grew. It looked dead – not a pretty sight.

Those two men, in particular, who had some knowledge of fitness training and who had helped many others in the camp, some less fortunate than me, did all they could to help me. Their most valuable contribution, it seemed, for months was hope. Although the arm continued to deteriorate they kept at me to keep trying to move it, or any part of it. What made me believe in them was the pure logic they came out with. Such as, if it hurts there is obviously feeling, if there is feeling it cannot die. If there is feeling then the nerves must still be connected, therefore it must be possible to move it – or parts of it – when the nerves have recovered sufficiently.

They told me to get a stick and try to hold it with both hands. I wrapped the dead fingers around the stick as best I could after much sweating and agony. On and off for weeks I spent hours trying to make those fingers curl on their own. I put all my willpower into that hand. All it gave back was pain.

Eventually it happened. The little finger moved. I had to stare hard to be sure the pain wasn't making me see things. I was holding the stick in my right hand, palm up, and it was lying across my left palm, the fingers not touching it. I was trying to squeeze the stick with both hands – as my friends had told me – trying to get 'sympathetic movement'. It worked, all those months of struggle paid off when I saw that slight twitch and realised it was in time with the squeeze with my right hand.

I rushed to tell the two PT instructors and found them together. They watched the finger perform, then told me not to rush things, but to work on it every day until I could move it enough to hold the stick. Then go back to them so they could show me the exercises which would build up the rest of the hand and arm. They agreed we had a great breakthrough, but cautioned against too much too soon. I did as I was told to the letter. Within a week I could touch the stick with my little finger and the finger next to it began to twitch on cue.

The next exercise involved the stick, a length of string and a small rock. I held the stick as before, with one end of the string tied to the middle of it and the rock tied to the other end. Then I rolled the stick with my right hand and gripped with my little finger – slowly at first – to roll the rock up to the stick. Soon after that great step forward, my shoulder began to work again. The elbow took longer to get going and, in fact, took several years to completely straighten.

The hand and arm put a little muscle and flesh on, but was still a scrawny claw after my release – although just before I was released the fingernails began to grow again.

It is likely that the weed helped in my recovery, although, by the time Poncho brought it, I was just about moving all joints except the forefinger, which took much longer to recover and was a pain unto itself. Being able to switch off the pain occasionally brought a great relief to my life, but I used the weed sparingly as I never knew how long it must last, or if there were any unknown side effects. After using it almost every day for about a month, I stopped using it for about two weeks to see if there was any craving for it. The only craving I got was for the relief from the incessant pain, so I felt secure with Poncho's painkiller.

The weed was not as tame as it is painted in modern life, however. Several experiences convinced me it could be lethal. One experience in particular is worthy of mention and underlines the previous use of the the word 'lethal'. Due to my crippled arm being permanently set in the shape it had been in the sling, it was easy to put things inside my jacket and under my arm without them being noticeable, so I was often asked to carry illegal odds and ends around the camp when the Chinese would have noticed other people and asked awkward questions.

One day, when I had just had a session with the weed, I was asked to collect a large, empty bottle from one point and move it to another. I didn't give it a second thought. Off I went and collected the bottle, put it inside my jacket, under the left arm, and set off. My route took me right past the Chinese Office and a couple of officers were sitting on the verandah steps. The usual guard was standing a few yards from them, eyeballing everything that moved. I thought it unwise to pass too close in front of the office and, with that thought in mind, decided to walk up on the opposite pavement. This would involve stepping over the rather deep monsoon ditch at the side of the road. Fifty yards from the office, I went towards the side of the road, this would look perfectly natural as I intended to look in my room as I passed and stop there a few seconds if the Chinese were showing too much interest.

At this stage I was quite normal, had made a logical appreciation of the situation and made plans to move accordingly – something which everyone does on streets anywhere. I approached the monsoon ditch with the intention of stepping over it as I had done hundreds of times before. I was two paces from the ditch, then it moved away. I hesitated, knowing it couldn't possibly, but there it was, just a ditch, only now it was about five paces away. I couldn't do anything stupid, like throw a rock to see how far away it was, without risk of attracting attention. So I sauntered on casually for a few yards then tried again. The ditch stayed where it should until, again, I was about one or two paces from it, then it moved again. I stopped, looked up and down the street. No one was taking any notice of me, the ditch, or anything else in particular.

I tried again. Same thing. By this time I was not only getting worried, but getting a bit close to the office to start doing anything stupid. So I turned around, strolled back down the road and had a go from a different angle. That bloody ditch bugged out again. There was an open patch between the huts and as I passed the spot I made my plan.

Turning again, I came back toward the open patch; just the ditch between me and open ground. I watched the Chinese, the guard had his back to me, the two on the verandah were chatting and laughing. I got the ditch lined up, ran and jumped as far as I could. When I took off I was convinced that I was about two or three feet from the ditch. I must have cleared ten or twelve feet – and landed smack in the ditch. The bottle under my jacket smashed against the rocks on the far side.

In the air I watched that bloody ditch keeping in front of me as it moved again. Jumping out of the ditch quickly, I glanced up the street, the Chinese were all looking, so I grinned and walked slowly towards my room. Everyone lost interest before I got to the room.

There were two or three of my room-mates in the squad room and they could see at once something was wrong. All gathered round and helped collect the glass as I slowly opened the jacket. There was very little blood, as I peeled off a heavy army pullover. There was then one piece of glass stuck to my ribs. I tried to brush it off. It wouldn't move, so I held it between finger and thumb and pulled. Then there was blood. It shot out like a tap turned on – all over the floor, and I had a long sliver of glass between my finger and thumb.

Someone jammed a towel against the wound, but it was obvious we needed real help. We went to the office across the street. One of the Chinese officers took us along to the medical room, which had not been opened long, and a Chinese nurse stitched the wound. Another Chinese officer wanted to know how it happened, so I said I fell in the monsoon ditch and there was a piece of glass on the side of the ditch. The officer who took us there said he saw me getting out of the ditch, so all was explained.

It was not the wound that bothered me. That ditch had appeared to move. After that I rarely used the marijuana, except at night. At night it was especially useful for a good sleep.

5 PRISON LIFE

The weeks and months dragged slowly by in the prison camp. Apart from our immediate worries of survival there was the background worry of what was happening in the outside world. News given us by the Chinese was, of course, heavily dosed with propaganda and consisted of grand denunciations of the 'American Imperialist Aggressors and their Running Dogs' who were hell-bent on sabotaging any peace proposals made by the 'Peace-Loving People of China and Korea' – as representatives of the 'Peace-Loving Peoples of the World'.

There was always the possibility that a Third World War would break out and we were constantly reminded by the Chinese that the first move in the event of global war would be the removal (by the Russians) of 'that American aircraft carrier off the coast of Europe'. Meaning, of course, the UK.

The war in Korea seemed to have no end in sight. We realised it could drag on for endless years. The possibility of never seeing our homes and families again was a stark reality.

The communist authorities tried hard to get prisoners to chant anti-American slogans. No one would. On one occasion we thought our friends in the American camp had cracked. A company of them were marching back to their camp, through our camp, from the lecture ground. They were all chanting, as they marched, 'We wanta Peace – We wanta Peace'. Then the punchline came – at full volume – 'We wanta piece of ass!' It gave us all a good laugh although, I must admit, a 'piece of ass' – in the American context – was pretty low on our list of priorities right then. Chinese–American relations, which looked on the up and up at the start of the chant, were noticed to hit a new low when the Commies eventually caught on.

Conditions improved gradually, with the Chinese trying to get the maximum Commie propaganda out of each step. The more they tried, the more they failed, of course, such is the British nature.

Every few weeks someone would 'go over the hill' – escape – but every time they would be dragged back within a few days to be jailed, having achieved very little.

One of the Commie gimmicks, which our captors tried hard to

instigate, was self-criticism. If anyone misbehaved, or was lazy, or did not do his job as well as he should, he was expected to stand in front of the assembled platoon or company and denounce himself in no uncertain terms. We were told self-criticism was the basis of good discipline and hard work throughout China, and all Chinese Forces in Korea practised it diligently.

Perhaps it was, and perhaps they did, but with us it was a dismal failure in its intended purpose, giving us more laughs than the Chinese thought good for us. Escapees were paraded to criticise themselves in front of the whole camp. One man in particular, who was away over the hill several times was paraded in the usual manner on the camp parade and lecture area. He was led up on to the stage by several guards. His head was bowed, chin on chest (most unusual for him), he looked decidedly roughed up. A long list of his wrongdoings was dramatically read out by an interpreter. Then he was pushed to the fore to criticise himself for his terrible behaviour. He stood for a few seconds, head still bowed. A great silence fell over the hundreds who had been paraded to hear his criticism.

Then he raised his face to us and said, in a strong loud voice, 'I bin a ba-a-ad boy!' The grin which then lit up his face put us all into convulsions of laughter. His popularity with the communist element hit an all-time low – for which he suffered – but that kind of morale booster was much treasured by the Brit and American trogs.

In early 1952 I had an idea to solve the great problem of where we were. All we knew was that we were in North Korea, about forty miles by road from another PoW camp at Pyoktong, which was on or near the Yalu River. The Yalu River forms part of the border between Korea and Manchuria, a province of China. The town we were in was called Chongsung, Chungsong or Chongsong. (There are a lot of towns with similar names in Korea.)

Part of the propaganda war was the issue of newspapers from China, including the English-language edition of the *Shanghai News*. In that paper there were often rough maps of Korea, showing the place of this Chinese victory or that UN atrocity. I copied those maps on to bits of scrap paper and gradually built up a fairly accurate master map which showed a lot of interesting things: roads, rivers, railways, heavily populated areas, the coastline and the distance between major obstacles, being just some of them. Other odd maps and things came my way, always adding to the total picture. There still remained the one great problem, our actual location.

It's all very well having a map of a country but it's not a lot of use if you have no idea where you are. Almost all the maps used to make the master map were quite legal in the camp as they came from the communist newspapers. The master map, and one or two others, however, were a different thing and had to remain well hidden. They were rolled very tightly and slid into holes in the wall of my room, plugged over and covered by two photos of my girlfriend, Ann; one photo of her leaning on a gate in Scotland, the other with her sister-in-law in the Cotswolds. (We still have the photos). So, we were getting maps together, but where the hell were we?

The propaganda kit came to my aid again. The *Shanghai News* was very precise in its reporting of 'Cowardly Indiscriminate Air Attacks by the American Imperialist Aggressors on Helpless Civilian Populations' anywhere and everywhere in Korea. The date, the place and – most importantly – the exact time of every air-raid was faithfully reported. So I became an air-raid watcher. Air attacks on distant towns were often visible at night. I watched and waited for the first glows and the sparkle of distant flak in the sky, then I logged the date, time and rough bearing.

To get the bearing, I had a spot near our room where I could lay a cross of sticks on the ground. This was lined up with the chimney of the next hut, which was in line from that spot to the Pole star. From knowing true north and looking over my cross towards the air attack, the rest was easy. The problem was the waiting. The *Shanghai News* arrived at our camp about a month late after publication on average, besides which we didn't get every copy. There was sometimes a gap of a week or more, sometimes only every other copy over a period of several weeks. So after each air-raid I had to wait a month before I could hope to match my date and time. But it worked, I fixed my position in Korea – and it was a long way from where a lot of people claimed it was. My position fix was confirmed by British Intelligence after my release, by which time I had confirmed it several times myself, with more fixes on distant air-raids.

During my air-raid watching activities I saw three or four aircraft shot down. The only raids I could get a fix on were at night, and it was often a lonely vigil on cold winter nights when no one with any sense was out of his room.

On one occasion I saw a bomber caught in a cone of searchlights. It was a long way off but I could see the aircraft diving and turning to escape. Then it seemed to give up and flew in a straight line to the south. I didn't see the night fighter, but the stream of tracer from its guns came from the darkness above and behind the bomber, leaving a flicker of fire on the aircraft. The fire gradually grew larger until the bomber dived steeply to the ground where the flash of its ending threw the western horizon of mountains into sharp relief. Other planes were hit by anti-aircraft fire from the ground defences and went to earth in a long, curving, flaming dive. The childhood experience of watching German raids on Birmingham and Coventry helped with judging the distance to the point where the raid was taking place, giving me more information for my fixes.

When we were first captured, we were casually interrogated a couple of times. An enemy who captures the whole damn unit, including the CO and the whole orderly room set-up, is not much bothered about interrogating ignorant little trogs. Once, somewhere before we reached the PoW camp, I faced an interrogator who (in retrospect) probably just wanted the practice. It was most likely at Halfway House, I'm not sure. He asked a lot of military questions about my unit. How long had I been in Korea? What towns in Korea had I visited? What other units had I seen?

He tried to be very sharp, brisk and vaguely threatening. It made me nervous because I couldn't see where it was leading. It didn't lead anywhere, just suddenly stopped after about an hour. I knew nothing anyway, having been in Korea such a short time before capture.

In the PoW camp there was at first a sort of 'documentation interrogation'. It seemed like they were building a picture for the records. Nothing much at all. Then Myrtle came. She was suddenly put in the platoon I was in, as platoon instructor. Her English was pretty good, she was aged about twenty-five to twenty-eight, five feet nothing in her heavy fur-lined army boots and had average to good looks. She had a reasonable sense of humour and could give and take with quite a bit of ribbing from the assembled platoon. Individually, she was very direct, correct and something else I cannot define. Rumour had it she claimed to have been a 'good-time girl' in Hong Kong. She didn't tell me that.

Myrtle had been with us for about two weeks when the problem started. One night after dark, one of the Chinese staff shoved open the door to our room and said 'Ladgah'. Most Chinese had problems with my name. When I sat up, he said 'office'. I pulled on my boots and followed him.

In the office there sat Myrtle, at a big table facing the door. She told me to sit on the bench on my side of the table, facing her. I had been to the office like this before, but in daylight, to face a man. This was different from the start. She said nothing. Thumbing through notes on the table in front of her, glancing up at me occasionally – for all the world like a very strict school teacher, looking through a terrible exam paper. Then suddenly after several minutes, 'What is your army number?' I told her. 'Why is it different from all the others?' I told her. 'Why are you lying to me?'

I said, 'I'm not lying, there must be other Brits with Regimental numbers here, ask them.'

Myrtle stared hard at me for a long time, then, 'You do not tell me what to do! You do not lie to me again.'

I think the number was an excuse. A point to start at. The questions went on, and on, and on. 'Where does your father work? Where does your mother work? Has your father got a car? Was your father in the army? Why did he make you join the army? Why did he not stop you from joining the army?'

I cannot remember all the things I was questioned about, or perhaps I should say I can't remember anything I was not questioned about! That first session went on for nearly two hours. The next night it started later and lasted about three hours. After a week I was dreading the night and that voice at the door, 'Ladgah, office'.

I was twenty-one years old, sitting across the table from a good-looking woman half the bloody night, and frightened to even smile out of turn. If I had something to hide I could have boxed clever (maybe). But I had nothing of use to them and couldn't understand why all this questioning. I thought it must be a kind of brainwashing to make me answer all questions automatically then suddenly they would pop the real one. But there wasn't a real one. How could there be?

Then Myrtle started on one of my friends and we took it almost in turn for a week or two. We never knew whose name would be called when that door opened. The worst stage of that episode came after three or four weeks. It wasn't the night she played 'kneesie' with me while she asked the questions – with her pet tommy-gun artist standing casually with that yawning muzzle an inch from my right ear. It was the night she did the kneesie and sent the guard off to get some cigarettes that I was certain there was someone else involved. She had sent the guard away for cigarettes before and I understood enough to know what she told him, but that was before the knees bit. She often gave me a cigarette during the calmer bits, when we were more like two people working on the same problem.

A few nights later it stopped. She never called us to the office again. We had both had similar experiences and were both at a loss for a reason. We were both convinced there was someone watching from the next office.

I was also involved with a couple of other interrogation sessions. There had been a few escape attempts and I think they were fishing for the next one. I was hauled into the office for no particular reason and had a nasty half-hour or so, full of questions about who were my friends in the camp and what friends had I got outside the camp? Then there was a sort of casual reference to the fact that he knew my parents' address and my girlfriend's address. Oh, and by the way, did I know how strong the

British Communist Party was? How dedicated they were to helping their comrades against the Imperialist Capitalist Aggressors? 'They probably have your parents' address, and I'm sure they will have your girlfriend's address. Now shall we start with the first question, all over again?'

Other men had the veiled threat thrown at them, but I doubt if the British Communist Party was aware of any of it, and I never heard of anything happening back home.

Getting out of the camp was no problem. It was quite easy to slip past the sentries in the night. The great problem was in moving across a well-populated country for hundreds of miles without being seen. After the Korean War I was told not one UN prisoner escaped from North Korea.

Twice I went up on the hill east of the camp to have a smoke in freedom. On both occasions the corn was high in the fields, giving plenty of cover at night. In retrospect it was stupid, but at the time it was great! As soon as we arrived at the camp, we were given a tobacco ration, for which we were truly thankful. A cigarette or a pipe can be a great help against boredom. The snag was paper. There was no paper with which to make cigarettes. I think we all tried to be pipe smokers at one time or another, but very few stuck to it. In my case it was more fun making the pipe than using it.

The rooms which had several layers of newspaper stuck to the walls were a boon. We peeled the walls and used that to roll smokes. Every bit of paper we could lay our hands on went up in smoke.

A lot of ingenuity was used to make darts and dart boards and many teams were formed. A darts league got under way – then the Chinese decided darts were a dangerous weapon and banned them. Packs of playing cards were made from old cardboard, bridge became a favourite pastime for many months.

Before darts were banned most of us belonged to one or another of the teams. The team I joined was called 'The Barley Mow', after a pub in Cheltenham. All the members were from Cheltenham area. The team captain, 'Robbo', was one of the great comedians of the camp and could be relied upon to knock a laugh out of any situation. His bright blue eyes almost always had a merry twinkle in them but, on occasion, he could make them look empty, vacant, wild. I don't think the Chinese knew quite what to make of him.

One game we played a lot was 'Aggis' or 'Haggis', I'm not sure which. This was like golf without the clubs. We set out a course of holes, got a pebble each and tossed it from hole to hole as per golf.

Once organised at the camp a couple of our people set themselves up as the camp's barbers. They cut hair and gave us a shave once in a while. Robbo started a craze which lasted for months. He had his head shaved 'down to the wood' except for four small patches, one back, one front and one each side above his ears. The four patches of black hair were in the shape of Heart, Club, Diamond and Spade. On another occasion he had his head shaved except for a six-inch pigtail which hung down the back and to which he tied a little bow of pink ribbon.

Another of Robbo's inventions was the Griff Hat. The kapok hats issued with our winter uniform had ear-flaps which were usually buttoned on top, but in very cold weather we let them hang down to cover our ears. Robbo fixed a piece of thread to one of his earflaps, so that, when pulled by his hand in his pocket the flap would rise up to uncover his ear – so he could better listen for griff (information). Or shit-house rumours more likely. Some of the guards were completely mystified by the Griff Hat, and possibly more than a little nervous of it, in combination with the vacant eyes.

After Robbo started the mad haircuts most of us followed suit, with many 'mohicans', 'hot cross buns', 'Friar Tucks' and a host of other weird and wonderful designs. It was all good for morale and raised many laughs.

Robbo was also one of the chief exponents of 'invisible sewing'. Several people were quite good at it. They would sit on their own with a garment on their lap and go through all the motions of sewing – thread the needle, tie off the ends, break the thread in their teeth – but they had no thread and no needle.

Another gimmick was table tennis. Two men would go through the motions of playing a fast game of table tennis, in an open area between the huts, or in the road. A crowd of prisoners would gather and all the heads would move together as the 'ball' went from end to end. The Chinese would notice the crowd and inevitably investigate. From behind the crowd it would look very realistic, but as soon as the Chinese tried to push through to see what was going on everyone would turn and move silently away. The Chinese were completely baffled and mystified by these things and probably thought we were all mad. (Not a bad diagnosis really!)

I have even seen two full teams, with attendant crowds of supporters, playing football, with no ball. There were one or two tug-of-war matches with no rope and, of course, invisible dogs. One or two went to great lengths to make dog leads and collars which really looked like an invisible dog was trotting along with them. They caused much laughter by stopping at the office steps and looking nonchalantly around while the 'dog' obviously did its thing on the steps.

Hole digging was said to be a good send-up in other camps, but it never gained popularity with us after the only time I saw it done, the instructor ordered the prisoners to dig it out again. The idea being to dig a hole then fill it in quickly as the Chinese approached, making them think something was buried there, so they would dig it out.

Some of my friends formed a band and singing group. They were all musically inclined and came up with the best music we were ever likely to hear in that place. One played drums quite well, using an upturned enamel bowl and an old tin, with a small brush for one drumstick. One or two played paper and comb. The star player was the bass guitarist, whose guitar consisted of a five-gallon oil drum with a cord attached to the handle and stretched to a broom handle, the other end of which was balanced on the rim of the oil drum. By increasing or decreasing the pull on the cord, higher or lower notes could be played. He became quite expert.

A couple of my friends from Cheltenham were good singers and became quite popular with their harmonised songs. Some men were poetically inclined and could remember poems they had learnt years before. Some wrote their own poems.

At Christmas 1952 the Chinese gave us each a little red notebook, ostensibly to take notes on lectures. On the front it had in English: 'Merry Christmas 1952,' and below that a copy of the dove of peace. The notebooks came in very handy for copying songs and poems which we then swapped around for reading. As we had nothing to read except communist propaganda the notebooks became a fund of reading and amusement, good for passing the time.

The air war was above us almost every day. American jet fighters would often plough almost the whole of a clear blue sky with their vapour trails. Great formations of them streaked through the sky, usually at a vast height, but occasionally a few would come down to lower altitudes. Two or three times, low enough to recognise the familiar white star markings. We envied them their war, with their coffee and doughnuts within the hour. Not even knowing of our boiled water and sorghum.

Aerial combat over the camp usually led to great speculation on all sides as to who had shot down whom. It was all too high up to make a good spectator event. On one occasion a jet fighter crashed a few miles away, the pilot coming down by parachute about a mile from the camp, up the road to the North. The Chinese were jubilant, it had to be an Imperialist aggressor. But when a truck rolled through the camp about half an hour later carrying a rather battered Asian man in flying kit there was a different opinion.

On three occasions the camp attracted the most unwelcome attentions of lone US medium bombers. These raids were at night. Twice we were bombed and once strafed. On the first bombing raid, I think three people were killed, one was a Chinese cook. The other two, strangely enough, were captured US aircrew officers. The second bombing raid didn't touch the camp, but the strafing run with a trigger happy belly-gunner rolling it out with a big .50-calibre came too close for comfort. All he really did was break a lot of roof tiles and scorch one man's back, but what a damn row he made about it! Several rounds hit our hut, but none entered our room. In the next hut a man sat up when he heard the shooting and a bullet scorched down his bare back and drilled a hole where he had been lying. In a camp full of soldiers he must have been the only one who didn't roll on to his stomach at the first shot.

One good story came out of the first bombing raid. A man had just left the latrine when he heard the bombs start to whistle. His soldier's instinct made him turn and dive back through the doorway where he collided with another and they both went in the pit.

During an air alert one night, when the bombers were going over very high up, we were lying in ditches around the camp, just in case. One of my friends had a watch with a bright luminous dial and an instructor shouted at him, 'So and so, put that watch out! Put it out at once!' Except for the panic in his voice I would have thought he was joking. He was deadly serious!

I think it was the same instructor on another air alert occasion, told us to be quiet and stop talking – as if the aircrews would hear us.

The day after the first bombing raid the Chinese got up a petition for everyone to sign. At least, I presume the Chinese thought it up, I could be wrong. The petition was to be sent to Kaesong, where the peace negotiations had recently commenced, to complain to the UN and accuse the US Air Force of deliberately bombing defenceless PoWs in a registered PoW camp.

There was some considerable argument between us and our captors over the extreme wording of the petition, which resulted in them changing the text quite a bit. There was also some considerable discussion between all the prisoners-of-war as to whether or not we should sign such a thing. At the time there was a lot of speculation about whether or not the free world knew of our existence. In the end most of us signed the petition. At least, we printed our name, rank and number on the sheet of paper, thereby (hopefully), letting the free world know that, at that date we were alive and PoWs in communist hands. It would also indicate our exact whereabouts in Korea. As far as most of us could see, the petition had considerably more going for us than for the communists – if it was ever presented in full to the UN negotiators at Kaesong. I never did hear if it reached them.

In the summer of 1952 there was a bacteriological warfare scare. The Chinese were quite convinced the Americans were dropping some dreaded lurgy on them. They had lots of photographs of the canisters used, some with flies crawling all around them. We were all inoculated against God knows what (tap water I expect) and a great hygiene campaign ensued. The announcement that the US had started this type of warfare nearly started a massacre in the camp. We were all rushed to the parade ground by the angriest, biggest contingent of guards we had ever seen. There were extra machine-guns around the parade ground and all the Chinese looked very upset – at us!

The camp commandant started the ball rolling going berserk in Chinese. We wondered what the hell we had done. Speculation ran rife as he ranted on. After a couple of minutes he stopped, there was a silence as the interpreter glared around at everyone to keep up the pressure, then he started: 'The American Capitalist Imperialist Aggressors have commenced to use Black Treacle Warfare!'

A deathly silence lasting two seconds, then a voice from the back of the British contingent shouted 'In barrels or baksheesh?' The whole camp dissolved into gales of laughter – and the Chinese went mad. More guards appeared and rushed up to the ranks of prisoners looking all set to shoot. The camp commandant looked fit to burst. It took a while to sort it all out, but the other interpreters present – the only Chinese who knew what was wrong – managed to get things under control before it became a massacre. A different interpreter took over and put us in the picture. There were yells of disbelief from the American prisoners, which got the guards on edge again. I was damn glad to get off the parade ground that day. But it was worth it!

A day or two later there was a swarm of big black flies down by the river. An American prisoner was found stuffing them into his mouth as fast as he could. Can't remember what happened to him – whether he went to jail, hospital or neither, but I do remember it upset the Chinese quite a bit. I think he was trying to prove he didn't believe the 'Black Treacle Warfare' story. He may have just been mentally deranged.

Another time there was a near riot when we had to organise a concert. There was to be singing, stories and poems. A few turns went fine at the start and the whole camp was on the parade ground as per lectures. Then one of our lot got up on the stage to read what was claimed to be a poem from the English classics. I don't know what he had on the sheet of paper, which had been vetted by the Chinese, but one thing is sure – it wasn't the poem he came out with! I can't remember how it went at the start but a few lines – the ones which made the shit hit the fan – I'll never forget.

They seek him here, they seek him there
They seek the bastard everywhere
Will he be shot or will he be hung?
That damned elusive Mao-Tse-Tung.

The camp went wild, but nothing compared with our captors. The great poet was bounced into jail so fast his feet never touched the ground. The camp was in uproar in his defence. A nasty looking situation was saved by Ding, one of our Chinese platoon leaders, whose English was almost nil. As one of the guards rushed up behind us, Ding snatched the tommy-gun from him, shouted 'office' (one of the few words he knew in English) and started off in the direction of our huts. We all fell in behind him and went back to our rooms. The concert was over.

The camp jail was a place to be avoided if at all possible. Our food was bad enough, but in jail it was much worse – and much less of it – if that is possible. The idea of jail, in most cases, was to get a confession in writing with a signature from the 'offender'. Very often the charges were absolutely ridiculous and men just would not sign confessions to rape and murder when all they had tried to do is get home. The jail in that prison camp was rough. It ranged from the cushy, which was standing or sitting in one position day and night for a week or two, with an ever alert guard to make sure you did, to hanging up by your handcuffs from a beam for long periods and, worst of all, I can only imagine the 'boxes'.

The boxes were just that. I only saw them once, from the outside, thank goodness. They were about three or four feet square, and made of wood. The only opening into the box – once you were in there – was a gap at floor level big enough to push a rice bowl through. (Three inches high by six inches wide).

Men were put in those boxes for two or three weeks. In at least one case it was for a month. They were handcuffed with their hands behind their backs. They usually had dysentery, but they never came out of the box or had their handcuffs removed for anything. They ate and drank like a dog. The guards kicked or banged the box to keep them awake – so they wouldn't miss too much of it by sleeping. There were variations on treatment in the boxes: some men were brought out once or twice per day to the latrine, some were not handcuffed, some had handcuffs with their hands in front of them. It varied, but it was all rough.

There were also cages. These were used same as the boxes, only the guards could watch you didn't get any sleep.

One man in particular, Derek Kinney, who made an absolute fetish of escaping and has since written a book, *The Wooden Boxes*, about his escapades and horrific treatment as a PoW in Korea, was given the full treatment at various times throughout his captivity, but never cracked. He was a member of the Royal Northumberland Fusiliers. I had met him after capture and knew him well by sight.

One incident I remember well but don't remember when it was, possibly late 1951 or spring of 1952, he had escaped and been captured again and we knew he was in solitary confinement. Some of us were sitting on the verandah outside our room when along came Kinney, looking rather roughed up but calmly nonchalant, as usual. 'Hello Lofty, how's the arm shaping up?' he said.

'Okay thanks,' I said. 'I thought you were in solitary again.'

He just grinned, 'I am, but they haven't missed me yet!'

He was so cool about it I thought he was joking.

Sure enough he wasn't joking, but he wasn't ready to make another run for it either, and I believe he got back into solitary before anyone found out. How the hell he did it I don't know. Kinney spent most of his captivity in solitary, sometimes in the boxes or cages, and at least once hanging by his handcuffs being beaten up.

Colonel Carne spent most of his captivity in solitary confinement too. I don't know what the excuse was, but I think it was because he would not conform. The political nuts tried to indoctrinate him and often attempted to make him admit to being wrong to serve in Korea. Someone who spent some time in the next cell to Colonel Carne told us something which cheered us considerably. One of the political nuts was yelling at Joe, over and over again, 'Have you realised your mistake?'

Joe Carne eventually answered. 'I am an officer in the British Army. I do not make mistakes!'

We didn't hear what the nut made of that one, but I would imagine he completely flipped.

While in captivity Colonel Carne carved a small cross from a lump of rock. After his release and return to the UK, Joe gave the cross to Gloucester Cathedral, where I believe it still is.

Another thorn in the side of Chinese efforts toward peace and quiet was our adjutant, Captain Farrar-Hockley. He was another escape-o-maniac, but, like young Kinney he only had to see a half chance and he was through. The problem with that method is you are always on the run in an even weaker state than necessary due to solitary confinement and almost nil rations.

I think the adjutant was the best chance we had of anyone escaping. He could navigate, had some knowledge of the area and where he was starting from, was well switched on mentally and had been physically very fit. Had he played it cool long enough to build up his resources and given himself a fair chance, I feel sure he would have cracked it. But, like

Kinney, he was a marked man from the start and had to escape sometimes from a prison within a prison. Farrar-Hockley's best attempt got him to within sight of the sea on the Korean west coast, and that was the route three of us from Cheltenham were intending to take in the autumn of 1953.

Once I had pinned down our position on the map and built up a reasonable picture of the geography of North Korea it was, in my case, a matter of waiting for my arm to improve enough. The plot was to move west, keeping to the mountains and more remote areas, moving only at night, hiding by day until we reached the coast, then find a boat and get over the horizon during darkness. The fruit in the mountains and crops in cultivated areas would feed us at that time of year. But it was not to be. Perhaps very luckily for me as, if recaptured and beaten up, I might have lost the use of my arm entirely.

There was not a lot of beating up of prisoners in jail, but it was not uncommon. Anyone who escaped from the camp was charged (on their return) with attempting to contact Imperialist agents, attempting to sabotage the Korean War effort, trespassing on the property of the Korean people and a whole catalogue of weird and wonderful intrigues to which they were required to sign a confession.

As an illustration of ridiculous charges, a friend of mine was talking to another prisoner, facing towards a path about seventy-five yards away, along which a Korean girl came walking. The Chinese guard standing a few yards away rushed up and hauled my friend away to the jail, where they charged him with attempted rape and for days tried to make him sign a confession. A lot of men said they would have been flattered to have been charged with rape at seventy-five yards.

Early in 1952 the brainwashing campaign was going full blast. To be able to do a better job the Chinese reorganised the prisoners into racial groups. We were moved nearer the town, into what had been the guards' quarters, thereby splitting us further from the Americans. All the black Americans were marched off to another camp miles away.

It was when the black prisoners were being sent off that the Chinese came up against the Brit at his best. We had one black trog in our ranks. Steve was a cook, attached to the Glosters, and was from St Helena. I was not aware of any colour prejudice in the British Army; I had never seen or heard of it, except with the Americans.

The Chinese swooped on Steve and were hauling him away to go with the blacks. The Brit contingent raised hell. There were shouts of 'He's a Brit, he belongs with us.' I remember seeing Steve's worried face turn into a big flashing grin when the Chinese hesitated and he saw we were all on his side.

Some of our people got a couple of the interpreters and explained that Steve was British and should stay with the British. Colour or creed had nothing to do with it if you were a Brit! The Chinese were thrown into confusion by all this – and several hundred prisoners there to prove it by yelling, 'We want Steve.' This went against all the commie patter about racism by the British Imperialists. In the end Steve stayed with us. The camp commandant changed the orders and Steve was classified as a Brit, irrespective of colour.

The Steve incident was one of those occasions when Chinese good commonsense overcame communist indoctrination and the PoW camp authorities, notably the camp commandant, used his head. These occasions were not as few as might be expected. They did Communism no good in our eyes as, without the stupidities of communism, most of the situations would never have arisen, but they did reinforce a belief, held by many of us, that the Chinese character would eventually triumph over the harsh regime in China.

Quite early on in our days at the prison camp, it was decided, by the Chinese, that flies were a menace to hygiene and we should do something about it. To solve the problem each man was given a fly quota. It started off at twenty flies per day, which had to be produced (dead of course) to be counted by the Chinese staff. The quota went up after a week or two, to thirty flies per day per man. After a while it became very difficult to fill the quota as flies became somewhat scarce. A simple enough solution to a vast problem. In spite of the original ridicule poured upon the idea by many of us, it worked. Flies went from being a big hygiene problem to being practically non-existent in about a month or six weeks. None of us minded killing flies, it was the tedious job of collecting and counting which we didn't like. I often wondered about the Chinese platoon leaders who had to count up one hundred men's fly quotas. Three thousand bloody flies to count every day; that's taking dedication to the limit! It is worth noting the guards and camp administration staff also had fly quotas to fill at that time.

From the day we arrived at the camp all drinking water was boiled. This was done in a big steel tub in the cookhouse, the same tub our sorghum or rice was cooked in. It was a sensible defence against the spread of disease and was one of the first instructions the Chinese gave on our arrival.

The British cooks would boil a tub of water about three or four times every day. When the pot was boiled one of them would shout, 'Hot water up!' and there was usually a mad dash for the cookhouse. Dozens of thirsty prisoners carrying our cans or bowls to be filled. There was always a madder dash when the food was ready, until we were issued with a couple of larger bowls to each squad, in which the squads rations were brought back to the room for equal distribution. The signal that the food was ready was always a call of 'Come and get it!'

All water in the camp came from wells around the streets, which had been the source of water for the townsfolk before we arrived. The well in my part of the camp was about thirty feet deep, and the task of lifting the water was helped by the use of a long pole which had the rope and well bucket attached to one end and a large section of concrete lashed to the other. It worked on the same principle as that employed by the barriers seen at factory entrances and the like, where a large barrier can easily be lifted by one person. Except that in the well's case the pole was upright when left alone and the bucket was lowered by pulling down on the rope.

One of our guards was nicknamed 'Goggles' as he always appeared on duty wearing a pair of fur-lined goggles. He was extra vigilant, screamed abuse and made a damn nuisance of himself at every opportunity. He carried a rifle, with bayonet always fixed and was very handy at making stabbing motions around anyone on the way to the latrines at night.

Goggles was out in the hills one day as part of the screen around a wood detail. A tree had fallen against some power lines and Goggles, seeing something out of the ordinary, had to use his bayonet as usual. He charged the tree and stuck the bayonet in it. I didn't see it happen, but those who did reckon Goggles went about twenty yards down the hill in a shower of sparks. A couple of lads from the RUR went and picked him up, recovered the rifle and bayonet, and took him down to a stream, where they commenced to clean him up and try to revive him. He was out cold for a bit, but came round eventually.

Other guards had seen this all happen from a distance and rushed to their comrade's aid. They brought him back to the guards' quarters, where he recovered enough to return to duty within a week or so.

The two prisoners who had shown compassion to another human being were highly praised by the Chinese, much to their embarrassment, and held up as an example to us all. Goggles probably had some brain damage, I don't know. After a few weeks on duty he had to be taken off guard duties. It was quite usual to see his unattended rifle (minus bayonet) leaning against the wall of a hut, while Goggles was inside playing cards with the boys. The Brits had helped him and were his friends; from now on we could do no wrong.

After the Americans were completely separated by distance and more sentries, we saw very little of them. Poncho came up one night to bring me a supply of weed. (Risking his neck in the process!) Most of our friends in the American camp we now saw only at a distance, at lectures on the parade ground, when we were under close supervision.

To a certain extent I missed the Americans. Some of the hardened old soldiers were good to listen to. Not only their quaint turn of speech – even for Americans – but they were always well up on shit-house rumours. If the war wasn't going to end next week, we were going to be liberated the week after, or maybe the week after that . . . Few of the Brits believed any of these rumours and I doubt the Americans were really fooling themselves, but it gave a good flavour to conversation. The only subject worthy of serious discussion was how soon we would get out of there.

Some of the older Americans had seen action in the Pacific, north-west Europe or Italy, and those men were much harder than most others from their country and they were good to talk to. As mentioned earlier the Americans had terrible morale problems – their 'three whole days and no candy' disease. From what I saw there were probably hundreds who died of just that.

At the other end of the scale there was a very young American who had fooled the army – or Marines, I can't remember which – into thinking he was two years older than he was. When captured, he had both legs trapped under a tank and had both legs sawn off at the knee without anaesthetic. He had been keen on boxing as a kid and we often saw him being carried by a friend and sparring with others. His morale was always sky high when I saw him. They told us he was still only sixteen years old when he was captured.

One of the real old sweats I knew pretty well, came from the American Deep South, his Southern drawl had to be heard to be believed. His remarks about those of his countrymen who, as he put it, 'Couldn't get on their back legs to piss with the wind,' were very much to the point. But he was the world's greatest optimist; the least thing in the real news – that given out by the Chinese, as opposed to the latrines – he could turn into another good reason we'd be out and home very shortly.

When General Mark Clark was made commander of UN Forces in Korea, the old sweat claimed he knew the General when he was just a 'Kiddy Cap'n'. He also claimed Mark Clark would have us liberated within a month. 'Lefty,' he said, 'that man is my kind of soldier! When he says shit – you squat and ask what colour.'

The US Army could have done with more men like that and less like the one I saw charge a very ill fellow American twenty dollars to fetch a rusty can of muddy water from the river, twenty yards away.

During the last few months of my captivity one or two of our people managed to get work in one of the Chinese offices, either typing or helping with translations. Most of us considered the situation decidedly suspicious so those involved were on very dicey ground. After a while the odd snippet of news from the outside world filtered into the camp and we found there was a radio in the Chinese office which, on the very rare occasions when they were left unguarded, our people tuned into the BBC Overseas Service. Hence the reason for getting the jobs.

It must have been very difficult to get any news, though, as the occasions of being left unguarded must have rarely coincided with a news bulletin. Nevertheless it was by this means that we first heard of the death of King George VI, several weeks after the event. The Chinese had not told us His Majesty had died because we were working-class people who would obviously have no interest in the relics of capitalist society – and in any case it may have disrupted our concentration on our political studies. An interesting combination of observations.

Any time we got the chance we would send up our Chinese political instructors. A few well-chosen words in the right place could often send them scurrying off to check the details with their books, or higher authority. Some of our people were quite good at it but the only time I remember getting them worried for a few minutes, it was more by accident than deliberate.

At that time the French were not doing so well with their Indo-China problem and we were forever being told of the great achievements of the 'Vietnam People's Army'. So a few of us from Cheltenham decided we must be the 'Chelt Nam People's Army', and I mentioned this mythical military formation within earshot of one of the Chinese instructors. He questioned me quite seriously about my connections with this undoubtedly great army, and I told him some of us were in it because that's where we came from, Chelt Nam.

It could have been a good send-up but another instructor became involved about half an hour later and he had obviously made a closer study of the Brit character, done his homework or whatever. The two instructors came to me and the more fluent English-speaking one said, 'Now, who has been mispronouncing the name of a town in Gloucestershire? You or my friend here?'

There were two main methods for the British prisoners to show dissent in the camp. When there was some problem and we were *en masse*, someone would start singing 'Land of Hope and Glory'. We found this really wound up the Chinese as they recognised it as a real Imperialist hymn. We often sang it when marching to or from the lectures. It can be very morale lifting, when the opposition think they have scored a point, to hear three or four hundred voices singing a song like that.

The other method was, as far as I know, home-grown in the camp.

Little hangman's nooses would appear as if by magic in all sorts of places; on the notice board; over the doorway to the Chinese Office; at each end of the office verandah, right under the noses of the guards. One of the sentries nearly went off his head when he saw a noose hanging from a tree opposite his sentry box. But he looked even more shaken when he returned to his sentry box and found one hanging over the doorway.

In the very savage winter of 1951–52, the cold was the worst I have ever encountered. While washing the cans in hot water the splashes would freeze on the backs of our hands. I know at least one sentry froze to death in his sentry box. Rumour had it there were seven or eight who froze to death at their posts that night.

Thanks to the Chinese we were well equipped to withstand the cold. The food had marginally improved by the time winter came, but the greatest life saver was the clothing with which we were issued. One or two rough cotton shirts to supplement our old army shirts, and a suit, hat, gloves and boots made of kapok. Quilted clothes similar to that worn by the Chinese, but not so well made, and blue instead of mustard colour.

We had by then been given a rush or bamboo mat for the floor of the room, and before the real cold came, we were issued with two or three heavy quilted eiderdowns which covered the whole squad at night. The heated floors of the huts did the rest. Without the issue of those things the Chinese would have had very few prisoners by the end of that winter.

I think it was in the late autumn of 1952, a very sorry-looking little group of bedraggled prisoners appeared on the road through the camp. There were about ten or fifteen of them. Their clothes were in tatters, some were barefoot. They had North Korean Army guards. In spite of the heavy threatening attitude of their guards, some of our people found there were a couple of Brits among them, and immediately rushed to the Chinese to request they do something about these poor scarecrows passing through.

The Chinese stopped the group, and there began a long parley with the Korean guards. Eventually the prisoners were handed over to the Chinese. I have a feeling the Koreans were glad to be rid of them.

One of the 'new' prisoners came to my squad, which had been reduced to eight or nine on the last reshuffle. They had had a pretty rough time as prisoners of the North Koreans and, I believe, had been more or less on the move ever since capture. The man who came to my squad was a Marine Commando whose small boat had been blown on to the enemy shore by a sudden storm after the engine failed. He was actually captured in his swimming trunks, which he was still wearing when I first saw him, several months after capture. He also had a piece of tatty blanket draped round his shoulders. Not the best gear with which to face a North Korean winter.

Conditions generally improved steadily throughout our captivity. By the end of 1951 we were getting steamed bread perhaps once every two or three weeks. This replaced the evening meal of sorghum. We were also getting a bowl of sloppy sorghum in the morning for breakfast. By late summer the greens were alternated occasionally with bean sprouts and later by actual beans. At first there was more water than beans, but gradually the ration increased.

In late Autumn 1951, maybe early winter, we had pork, for one meal only. One average-size pig was killed, chopped up and boiled in a lot of water to feed about five hundred men. Anyone who got more than one tiny bit of meat was lucky. Nevertheless, the water tasted bloody marvellous.

At Christmas the Chinese put on extra rations, we had pork again, at about the same ratio as before, and chicken as well. I think we had chicken once or twice before Christmas 1951. Six or seven chickens, plucked, chopped up (bones and all) and boiled in a lot of water to feed about five hundred men. Again the water was heavenly. Rotten fish appeared once or twice, but even we starved critters had problems with that.

Gradually, through 1952, the food improved. Pork and chicken came more often and more of it. Steamed bread became a weekly event by late 1952. Sorghum began to be replaced by rice. Millet disappeared altogether. Greens were occasional, having been replaced by soya beans by late 1952.

Sometime in late 1952, perhaps at Christmas, we were each issued with a toothbrush, toothpaste, soap and towel. These items were very welcome and raised morale as well as our general hygiene standard.

On two or three occasions, maybe Christmas 1951, May Day 1952 and Christmas 1952, our captors surprised us by issuing every man with a good tot of wine. They called it 'peace wine'. It amounted to somewhere between one and two tots, as per spirits, and was just about raw alcohol. One or two people still had their petrol-fuelled cigarette lighters, and found they went well on 'peace wine', so it was strong all right. Had we been given about five tots each there wouldn't have been much peace in that area. Some of us called it firewater.

By early 1953 we were being fed pretty well by previous standards. No one should get the idea we were being fattened up though, because although by the time of my release I had been on comparatively marvellous rations for two or three months, I was considerably underweight, only nine stone ten pounds. Normal food was hard going when I was first released, but in two days I put on eight pounds.

Reports in communist newspapers in the United Kingdom and elsewhere, played up the pork and chicken we received so that their readers were very likely given the impression we were living the life of Riley. At least two communists came to Korea to see for themselves. One was a Brit, a woman; the other, a male, was Australian. The Aussie brought us a real football and was told what to do with it – inflated! By that time the Chinese had already produced a couple of footballs for us and gave us a certain amount of freedom to play matches on the big parade ground. So, telling the traitor commie what to do with his ball was no problem.

The fact that Commies from back home could come and wander around in North Korea while British troops were fighting and dying with the Commonwealth Division to the south, struck us as significant. It meant there was no difference between our Commies and the rest of them, anywhere in the world.

In our minds British Commies took on a different aspect. They obviously condoned stabbing Poland in the back, the slaughter of the Polish Officer Corps at Katyn Wood, the Siberian Salt Mines for political dissenters, the wholesale slaughter of all who would not conform to the ideals of Lenin, Marx and the others. More to the point they condoned North Korea's invasion of South Korea in 1950, and the ensuing slaughter, which swept Korea from end to end.

We did not hate the Chinese. They were people like us, who were trying to do their job the best they could. They were fighting for their country and were respected for it. But almost every British prisoner hated the system which had caused the war by trying to force itself on other people. We respected the Chinese much as we had respected the German troops during the Second World War, but communism was hated as equally as fascism.

Dictatorship by one man, or dictatorship by one party, what's the difference? Both become a law unto themselves, trampling on the laws of their own country – before starting on someone else's. The methods become irrelevant so long as the end result is achieved: total subjugation of the masses.

The traitors from the UK who came snivelling around in Korea should have been accorded the same treatment they would have received had it been the other way round.

Can you imagine the welcome home party if a Chinese journalist or politician had hobnobbed with the Americans in South Korea? Talked to Chinese prisoners-of-war, told them they had been liberated from a dictatorship, then gone home to tell how well treated they were. Do you think there would have been a newspaper in Peking which would have dared employ such a person – let alone publish anything they wrote? These British Commies do not think past their noses to see they are only free to hobnob with the enemy because their home is in a democracy. Their attitude is to say Capitalism is too weak to oppose them – as they represent the people.

We may not have a salt mine patch to send them to but we do have a little place called Rockall. If they are not bothered about the means – the end result is the same.

6 REPATRIATION

In March 1953, having been called to the office, I was surprised to be given a thorough medical examination by a Chinese doctor. He spent a lot of time examining my arm and shoulder, making notes on a pad every few minutes. The examination lasted about half an hour, after which I was told to return to my room – without explanation.

Medical facilities at the camp had improved somewhat over the time we had been there. At first there was nothing. Then, several weeks after we arrived, a Chinese woman doctor appeared. She could speak fairly good English as she had studied in Edinburgh. She was really a surgeon but could do very little for us as she had little or no surgical equipment. Her only medicine as far as I know was iodine. It was used for everything – even toothache. I don't remember when she left, or what the medical situation was for the last few months. Men still died occasionally in the American camp, but the burial party was a comparative rarity on Boot Hill.

In mid-April 1953 I was again called to the office. This time they told me to report back in one hour with my kit as I would be going into hospital. There was a mad rush around the camp to say goodbye to my friends, some of whom gave me their home addresses on a tobacco packet. Rumour had been rife that there was going to be an exchange of disabled prisoners so we were not completely taken by surprise as everyone seemed to think I would be a prime candidate. I personally thought that in my case it was more likely the Chinese would give me the treatment required to cross me off the disabled list, if they qualified me for anything at all. The Chinese told me nothing, except where to go and what to do.

Arriving at the hospital, the same old Japanese temple on the hill, where so many had died when it was just a death house, I found it was now a real hospital with proper beds, real nurses, doctors, the whole works. An English-speaking Chinese officer told me they would operate that night to remove a bullet from my ribs. A feeling of doubt crept into my mind but the familiar feel of a clean, efficient hospital was reassuring. Late in the evening a Chinese man, who turned out to be the surgeon, came to examine my arm and ribs. He seemed efficient and had the confident air of a man who knew what he was doing. I felt further reassured. About midnight, a nurse came and escorted me to a small operating theatre. All the usual preparations were made, a nurse told me the operation would be completed with only a local anaesthetic. For some reason I felt relieved they were not going to put me out.

All went well until the surgeon obviously found the bullet was jammed into bone and he had to get some bigger tools to cope. That's when the air-raid siren sounded and all the lights went out. About two seconds later two of the nurses had hand torches going – as if they had expected the air raid. The surgeon worked on steadily, while the bombers droned overhead and his patient sweated blood while thinking what a stupid situation to be caught in if those bombers decided to unload.

Listening to that menacing rumble and roar in the sky above, waiting for the first whistle of the bombs, deciding which way to roll off the operating table must have passed the time quickly. Before I knew it the surgeon had stitched up the wound, handing me a very battered tracer bullet and looking pleased with his handiwork.

I hardly realised I'd been holding one of the nurses' hands until I reached for the bullet and she was laughing and rubbing her crushed fingers.

One of the nurses said, 'Okay, go back to your bed. We will look at it again in the morning.' The bombers were distant, and soon after I reached my bed the 'All Clear' sounded.

Early in the morning the surgeon and a few other Chinese doctors and nurses came to check the site of the operation. They seemed well satisfied and left. Half an hour later I was cleaning my teeth in the yard when a nurse called me and told me to go at once to the gate at the main entrance. My first thought was to put away my washing kit, but the nurse stopped me and said, 'I will take that. You must go to the gate at once.'

At the gate a Chinese officer checked my name off a list and told me to go out of the gate to sit with some other prisoners on the steep hillside in front of the Temple Hospital. There were several other prisoners, British and American, already sitting on the grass, no one knew what was going on. A few more men came out to join us until after five or ten minutes there were about twenty or thirty of us sitting on the hillside, which was stepped at about two-foot intervals.

A Chinese officer walked to a position below and in front of us. He looked us all over carefully, spoke briefly to other Chinese hovering nearby then smiled and addressed the assembled prisoners. 'Gentlemen, there has been an agreement to exchange sick and wounded prisoners. You are going home.'

There then started a dream-like, confused, hazy few days which I would like to remember more clearly. Some of it, that is. First of all, I didn't feel excited, being convinced it was all just another dream. The whistles would blow any minute, awakening me to another dawn, another roll-call, another lecture, probably sorghum and beans. In any case, after all the months of haggling and mind changing at the peace negotiations, it was just as likely they would all fall out and cancel the lot by lunchtime.

We were not allowed back into the hospital. Our kit was brought to us in the wooden huts we were taken to. There were some speeches by the Chinese, they were short and all except the first – from the camp commandant – were in English. There was a complete lack of propaganda patter, the main theme of the speeches being to wish us and our families health and happiness in the future. It was one of the few times our captors spoke to us with complete sincerity. I am quite sure those Chinese who spoke to us were happy for us to be going home.

We were given a quick meal with strong Chinese wine to wash it down, taken out to open trucks, loaded, checked and away we went. The little convoy of three or four trucks rumbled northwards through the prison camp, but I was too disbelieving, too upset at having to leave all my friends – who stood at the roadside – to do more than wave weakly as we passed them.

Once clear of the camp we gathered speed and made good time along a winding road through wooded mountains to a prison camp at, I believe, Pyoktong, close to the Yalu River. More trucks loaded with prisoners joined us and, after an hour or two, we were on the road again with a much bigger convoy heading south and west until night.

At night we stopped somewhere, I've no idea where. Memories of that journey are dominated by pain. Where we stopped or even how many days we travelled is all lost in a haze of pain. Strong red wine helped deaden the pain but it deadened the memory as well. The roads were dirt roads, a bit the worse for wear and showing plenty of signs of the continual bombing. They were full of holes and humps, and the trucks went like hell. Every morning I burst my stitches and bled all over the place. Every night they stitched me up again, without anaesthetic. The operation wound and the bleeding didn't bother me much but the continual bouncing and jolting played hell with my shoulder and arm. There were no seats in the truck. At times we were all in a heap on the floor, trying to find something to hang on to. Obviously there was a deadline to meet, as some of the Chinese were concerned for our well being, but said we had to move quickly.

At last we came to a place with a bigger building than usual, full of bunk beds. We got ourselves cleaned up. My blood-soaked clothes were replaced. We were thoroughly searched from head to toe. All our kit was searched and I noticed with horror that the tobacco packet with my friends home addresses on it was now a tobacco packet with part of an escape map on it. I had checked my kit at the hospital – and destroyed the wrong tobacco packet!

The Chinese searching my kit looked at the rough map, glanced up at me then put it with the rest of my gear. I died a thousand deaths while he finished the search, then he waved me on to join the others.

The moment of release came. My name was called in the last batch of prisoners. It all felt like a dream with the hated whistle just one breath away. Wondering if it was all a trick which would finish up in some horrific punishment cell for having that map. I think we travelled by bus to the exchange point, then were checked again, and walked over to a British ambulance where the familiar uniforms and well-fed Brits helped us in. The doors banged shut then off we went.

As the ambulance moved steadily along a dirt road, across an open, grassy area I noticed through the window to the left a hill, with three or four plumes of smoke rising near the top. With some shock I realised the plumes of white smoke were made by exploding phosphorous shells. To the right of the road, as we were climbing a slight incline, there were a lot of Chinese troops, moving forward on their stomachs, almost flat in the long grass.

We went over the top of the rise and about one hundred yards down the other side met a platoon of US infantry, walking nonchalantly up the side of the road in single file with their weapons slung on their shoulders. At that particular moment I definitely did not feel out of the woods. The two or three British officers in the ambulance didn't seem at all bothered by the situation outside. It worried me a lot, but I said nothing.

The ambulance ride was a short one, perhaps ten or fifteen minutes before we arrived at Freedom Village, a set-up especially prepared to receive repatriated prisoners-of-war. A very pretty young lady in the uniform of the British Red Cross met me from the ambulance and took me to a large room or tent where there was a very long bar with several barmen standing behind it, all looking eager to serve up a vast array of drinks. There were perhaps half a dozen people in the room. None of them were ex-PoWs.

One of the barmen said, 'What would you like? Name any drink you fancy.'

The Red Cross girl said, 'They claim they can serve any drink in the world at this place. Go on, try them.'

The whistle must surely blow now. I said, 'Banana milk-shake, please.' And it likely took all of three or four seconds for a huge banana milk-shake to land in front of me. I remember looking at it in amazement.

Several people were laughing, I may have been one of them.

I don't remember seeing any other ex-prisoners at Freedom Village. I had been released with a batch of mostly Americans, also, somewhere along the way my dressing had been changed, the site of the recent operation had been checked by a British medical officer, who had nodded his approval at the work of the Chinese surgeon. I had also been weighed and was rather surprised to find I only weighed nine stone ten pounds. I didn't feel that skinny. Before capture I had been roughly fifteen stone eight pounds. By the time I arrived at the bar the others had, I believe, been flown out by helicopter.

The Red Cross girl sat with me while I sipped that marvellous milk-shake. She asked my name and unit, then for the address of my next of kin. Afterwards keeping up a quiet, light-hearted patter, which was great, all about the bar and Freedom Village. I didn't feel so alone and out of touch.

After about ten minutes the military machine took over and took me to a room where I answered a lot of questions, then to another room where I got rid of my prison uniform and put on a British combat suit for the first time.

The Red Cross girl did her job well. My parents were glad to hear I had been released but very worried that I had been released in an exchange of wounded prisoners. They had hardly time to get worried before receiving a telegram from the Red Cross girl telling them she had seen me and I was okay. She will never know how much we appreciated that telegram.

Somewhere along the way we were told twenty-two British prisoners considered to be unfit for further military service had been exchanged for over seven hundred Chinese prisoners. Someone said, 'Now you can work out what you are worth in Chinese.'

I asked when they thought the rest of the prisoners would be released. 'Not until the war is over.' My thoughts kept going back to the camp, the friends I had left behind. Anyway, that bloody whistle would end all this rubbish in a minute. The prison camp was easier to believe than all I was now confronted with. It got worse.

I was the last British ex-prisoner to be moved from Freedom Village to hospital. The whole United Nations press had waited days – or perhaps weeks – to get an interview with us at Freedom Village. A British officer put me in the picture about the press. He said all the other British ex-prisoners had refused a press interview and were now gone. I was the last hope as there would be no press interviews in hospital. The decision was mine entirely. Knowing nothing of press interviews or the world of international news, contact with the press at that time was pretty low on my list of priorities. I didn't relish the thought but for some stupid reason I thought of the problems the press were having and said okay, I would do it.

Two British Intelligence officers briefed me on how to cope. They reminded me – as if I needed reminding – there were still a lot of men in communist hands so be careful what I said. We agreed between us that any questions which shouldn't be answered they would tap my leg under the table, all I had to do was wait for the next question. They would sit either side of me and fend off the more stupid questions openly.

We went into the press interview room to be immediately blinded by scores of flashbulbs. There were a hell of a lot of people there, all banked up like in a sports stadium. I kept the answers short, saying as little as possible. I don't know what my reactions were supposed to mean, but I sensed a lot of people were interested. After a few minutes the Intelligence people got me out of there. The next thing I remember is shaking hands with the Red Cross girl then getting into a helicopter. The chopper took me to a military hospital in Seoul, where I got another shower, my dressing changed and a clean, soft bed.

There was food, but I don't remember what it was, or even where I had my first good food for two years. After all those months of longing for real food, when I got it I couldn't eat it. A few bites of anything and I felt full up.

We were in an American military hospital. In the same ward there were American ex-prisoners who had been released that day. The last thing I remember hearing before sleep crept over me was a whining American voice from the other side of the ward telling a crowd of people how he had seen hundreds of American prisoners driven over a cliff to their deaths, by the Chinese or North Koreans.

I remember thinking: 'This is where the bullshit begins – if that whistle don't blow.' It was 21 April 1953.

The next morning, no whistle, but we arose early, got cleaned up, went to an airfield and boarded a Dakota. I don't remember how long the flight was. It was great to be really covering some miles without a great increase in pain. The pain in my shoulder and hand was still a big ache, but with my great faith in medical science I thought our surgeons would cure it with a few deft cuts in the right places. I can't say I was looking forward to the waiting time. In fact, I expected they would operate in Japan.

We landed somewhere in Japan and were taken at once to the British Commonwealth Military Hospital at Kure. Once there, we came under Commonwealth control completely. I began to believe that I was really free. The hospital staff were very efficient and very humane, not always an easy combination.

Commonwealth set-ups always struck me that way, a sort of easy-going super efficiency. Men who served with the 1st Commonwealth Division in Korea said that feeling went right through the division. At soldier level it gives a great feeling of confidence. It has been said that 1st Commonwealth Division was one of the best divisions ever put in the field, anywhere. Its record shows that is probably true.

Just before my release I met a young soldier, Pete, who had been captured only six months previously. He was in the Welsh Regiment and had been blinded when captured. His home was not far from mine, in the Cotswolds. Pete was full of how good the Commonwealth Division was. He made me feel I had missed out by not having been part of it. Other people we met gave the same impression. I had never known such pride in a division before.

In Kure, we were all given a thorough going over by the army doctors. I soon realised there was not going to be a quick solution to my crippled arm. The best advice they could give was carry on as before. It seems they thought I had done a damn good job so far. A specialist who examined me said if I'd had the same injuries in England and gone straight into hospital, my chances of living would only have been fifty-fifty. The fact that I had not only survived under adverse conditions but had managed to recover a good deal of movement in the arm surprised them.

The X-rays they took showed I was still carrying eighteen pieces of shrapnel in my chest, arms and back, but no sign of the other bullet. (There was a note on my medical records, query plastic bullet?) It has never turned up.

The shrapnel was not worth digging out. That was easy to accept as, without X-rays, I would never have known it was there. Some had come out in boils during captivity, a little has come out since. The hardest thing to accept was the thought of being crippled. My arm was so weak, thin and almost useless. I had thought an operation or two would fix it. I was assured that good food and time, plus continual exercising would very likely give me a reasonable arm – eventually. One of the doctors made my two years' imprisonment seem worthwhile when he said the arm would almost certainly have been amputated if I had not been captured.

It was obvious that food would soon make a difference as only two or three days after release I had put on eight pounds.

By the time we had been in Kure a day or two it was beginning to sink in that we had not done so bad at the Imjin River after all. The Glosters were now famous for holding out long enough to delay the main Chinese offensive at that time. Stopping them for an extra day or two from using the road to move their guns and supplies south gave the UN time to form a stop line which crumbled the whole Chinese spearhead, and our people were back on the Imjin within a month. Nevertheless, the feeling of failure, which had hung over me for two years, persisted.

Later, much later, I learned that 29 Brigade, on 22 April 1951, had six Chinese Divisions, about fifty-four thousand men, all to themselves. In fact, because of the importance of the road through our area the Glosters were elected for the main thrust of the Chinese attack and were hit first by three divisions of the Chinese 63rd Army (twenty-seven thousand men) with elements of the 64th Army to our west. The 63rd Army is estimated to have suffered over eleven thousand casualties by the morning of 25 April 1951 and was replaced by the 65th Army whose job was to finish off the Glosters, then advance south as fast as possible. It is worth noting the Glosters, thanks to heavy reinforcements, were operational again within a couple of weeks, whereas the Chinese 63rd Army was withdrawn back to China and did not take further action in the Korean War.

The feeling of being a failure diminished. The resentment of those clowns who sent us into action with outdated weapons against those sort of odds – well, enough said.

We were not allowed to leave the hospital while in Kure, so the nurses volunteered to do any shopping for us. I asked my nurse to get me a good camera so I could take photos on the way home. She got me a very good camera and several rolls of film which I used up rapidly over the next week.

After about three days at Kure, we boarded a Royal Air Force Hastings for our flight back to the UK. On doctor's orders our limit was eight hours' flying per day. That meant it would take a week at least to get home.

Our first stop was at Clarke Field, a US Air Force Base near Manila, in the Philippines. At Clarke Field, I insisted on having a tooth out. It gave me a dog's life every time we got airborne, so out it came. At least that was a pain which I could do something about.

The Hastings we were in was adapted for casualty evacuation. We had four or five stretcher cases aboard, two or three of which were men who had recently been wounded and not ex-prisoners. There was plenty of room in the aircraft – it was not even half full. Next stop was Singapore, where we again spent the night in hospital, then on to Colombo, Sri Lanka, the next morning. I remember nearly making myself sick on fresh pineapple in the hospital at Colombo. From Sri Lanka to India, then on to Habbaniyah in Iraq where the Air Force base was very proud of its beautiful gardens in the middle of a desert.

Our last night's stop was in Malta, where we each received a welcome home letter from Lady Mountbatten and an apology for not being there to greet us, as she had to leave the island just before we arrived. All the way home the forces did a great job of keeping the press away from us. We were really well looked after. There were many doctors, nurses, orderlies and cooks, whose day to day work touched our lives at that edgy time, who will never know how by just being normal and doing their job helped tremendously with a difficult adjustment in our minds. Little did they know I was now awaiting that whistle with ever-increasing dread and daily increasing certainty that it would blow any minute. I don't know if any of the others had the same things in their minds. I was afraid to mention it for fear of looking stupid – or giving some clown the idea of blowing a whistle.

Eventually the green fields of England were below us, one of the best sights I have ever seen in my life. We landed at RAF Lyneham, in Wiltshire, where my parents and sister were waiting to meet me off the aircraft.

We are not an emotional family, so it was just a matter of 'Hi, Maw, hi, Paw, hi, Jan,' and I was home. It was great to see them there, the happy faces and words of welcome and Mother's obvious relief that I wasn't crippled: 'They'll fix all that in a week or two.' The little sister I had left behind nearly three years before had grown a foot or so, and now looked almost grown up; I didn't recognise her at first.

Half an hour later we went our separate ways. They went home and I went with the others to the RAF hospital at Wroughton near Swindon for the night. The next day we were taken by ambulance to the Army Hospital at Tidworth, on Salisbury Plain, to get our final check-up and documentation before going on leave. I think we spent the night at Tidworth, then (each with a nurse as escort) we were sent home.

Those of us going to Cheltenham area, about four or five, went by train from Swindon. I remember looking from the train when we were on an embankment and for the first time seeing a mass of television aerials. I asked someone what they were and we all stared in amazement at these strange shapes sprouting up from all the houses. The nurses thought this was hilarious and we all finished up getting probably the best laugh we've ever got from television.

At Cheltenham Station we had a problem. The nurses were all going home for the weekend once they had delivered us to our families, but as we all were catching taxis home from the station, this seemed to be taking duty too far. Except in the case of Pete who, being blind, we allowed had a problem. He didn't see it that way, however, and insisted there was no need for his nurse to ride out to the sticks and back when he and a taxi driver could do the job just as well.

The problem was solved when Pete somehow sneaked into a taxi and went past us shouting 'goodbye' and 'tally-ho'. The way he took off there was no way we were giving chase, so we said many thanks and cheerio to the nurses and followed Pete's example.

So, I arrived home. The walk up the garden path, Mother's cooking, strong homemade wine – the whole scene of being really home.

Ann was there, just as I remembered her, to make the homecoming complete. The physical move from prison camp to home had taken less than three weeks. The mental move took longer. Every morning for a week or more I awoke in a cold sweat – thinking it was that bloody whistle that awoke me.

The taxi driver who took me home wouldn't take any money from me, said it was worth every penny to see me get there. A twelve-mile taxi ride for nothing. How did he know what I'd been through? I had sat in the back and hardly said a word all the way, just gave him directions.

I soon found out. My picture was in all the local newspapers and probably some of the national ones as well. My stupid grin was all over the cinema newsreels and on television.

There were nine hundred British prisoners in communist hands when we were released and we twenty-two were supposed to be the worst cases. 'Unfit for further military service.' So we were news.

For several weeks I found it difficult to pay for anything. Cinemas, buses, taxis, pubs, they all recognised me and refused payment. Everyone wanted to buy me a drink, but I couldn't take it. Half a pint was over my limit, any more and it wouldn't have stayed down. I tried a few times – not that I was fond of beer anyway, just to be sociable – and suffered for it. I never went in a pub willingly, always press-ganged by a well-meaning crowd. People I'd never seen in my life before.

I felt I was stealing human kindness that was more deserved by others still rotting in those prison camps, so I shunned the towns and stayed at home. I mostly just walked and ran through the woods, went shooting on my own, or sometimes with my father, when he had time. I walked miles through the woods and along country lanes with Ann, when she could be there. Her sense of humour and light-hearted view of life in general did much to restore normality.

After a couple of weeks, a letter came from a branch of the Intelligence Service, saying they 'would like to have a chat'. They offered me the choice of going to London or an officer coming to my home. When the officer came he was well pleased at the chance to get out of town and appreciated my reasons too. I told him where I thought my prison camp was. His maps confirmed my position fixes, much to my satisfaction. Then he showed me good air photos of the camp and I was able to fill in a lot of detail about what was what.

He visited my home several days for a couple or three hours at a time. On one occasion I asked his advice about letters I had received from people whose relatives were still classified as 'missing'. Most I had never heard of, but a few I had seen and heard dying. There was no way I could do anything, I couldn't tell people a load of rubbish – neither could I tell them the truth. Some of the memories were too fresh to even try to think of – or connect to real living people. When people came to the house my parents told them I was away.

The officer took note of the men I knew were dead, but there were no official guidelines as to how to cope with the problem. He said I'd have to rely on commonsense, which had done okay so far. During our conversations it became clear to me that the UN Command had kept quite close tabs on the prison camps and what happened in them. If only we had known of their interest during that two years of captivity, the effect on morale would have been terrific. One tends to feel very much cut off from the outside world when being a PoW, and without the slightest indication that your own military are interested, it feels like they don't want to know. I think I mentioned to the Intelligence officer at the time, it would do an absolute power of good if one of the seemingly hundreds of jet fighters forever over our camp could drop down for a quick circle around at low altitude or, better still, a victory roll right over the camp at about one thousand feet! Or perhaps not a victory roll. The shithouse rumour brigade would be demanding the weapons from their guards, insisting it was all over and we had won!

All in all the officer's visits did me a power of good. At least I knew the men in the camps were not forgotten and never had been.

My family and Ann's family did a great job of building me up and I put on weight rapidly. The shoulder and hand were still badly wasted, weak and painful, but in a happier environment it was easier to put the pain from my mind for ever longer periods. I concentrated on getting fit and controlling the pain so I could build up the strength in the injured arm. I found it was useless to exercise the arm too much, as the pain became intense. The arm became uncontrollable and almost useless afterwards. So I had to be patient and move things on very slowly.

My leave was for an indefinite period. I was on double rations (ration cards, as per the Second World War, were still required for some food) and double ration pay. It helped – with my appetite!

I split my time between my home and Ann's home and, to help with the travel problem, I bought a second-hand motorcycle for eighty-five pounds: a 350cc BSA. The only problem was the clutch control which, being on the left handlebar was meant to be pulled towards the handlebar by the fingers of the left hand being closed. I could not work the clutch in the normal way, but I managed to use the clutch by curling my fingers around the handle and pulling back with the whole arm. Once I got used to the clutch it was good exercise for the arm. After a few months I had built enough strength into the hand and fingers to use the clutch properly.

The people of Guiting Power, the nearest village to my home, and people from surrounding countryside had a collection for me and raised fifty pounds, quite a lot of cash in those days, which they presented to me at a little ceremony in the village hall. I was very grateful for the cash, but very surprised that they should think of doing it. There was no way I could refuse the good folks' gift. But again I was being given something for only doing the job I was paid for. I knew quite well what I was letting myself in for when I joined the army. The way I saw it, whatever I had to suffer was my own doing, part of the job. In fact, I felt I owed it to these people, and to the army, to get back to full fitness as soon as possible.

Nevertheless, many years later, in what has become another world, I thought back to those good folk of Gloucestershire, their honest, undemonstrative but sincere welcome home (no questions, no hassle, just 'Good to see you home, boy'), when I read reports of the treatment meted out to some of the American forces on their return home from Vietnam.

Some US Armed Forces were met at airports and docks by hostile, spitting, jeering crowds of fellow Americans, which is drastically wrong. It takes all sorts to make a world, and people will voice their opinions but there must have been something lacking in the make-up of the authorities who would allow their men to be subjected to such provocation.

Those men were wearing the uniform of their country. Many of them had thrown their lives on the line and were likely surprised to have survived. Theirs not to reason why, they had done their job as I had done mine many years before. They too deserved, at the very least, the respect of those who had sent them.

As part of my arm-strengthening campaign I made myself a bow and some arrows from local materials, as I had done in my younger days. Even when still at school I had made bows and arrows which would go right through a five-gallon oil drum at about fifteen paces. At first I had to use a very light bow, but as the arm gained strength the bows became stronger and eventually I could send the arrows nearly two hundred yards.

The countryside and woods are the best gymnasium in the world if, like me, you are not put off by rain, wet ground and mud. Running a lot caused considerable pain at first, but as the arm grew stronger, so the pain diminished until it was no more painful to run than to stand still.

Apart from regular exercises, I used the arm for anything and everything I could, such as opening doors and playing darts. One thing I could never master was lifting a cup to drink. The arm would tremble too much to control when in the drinking position.

After I had been home for a couple of months, I returned to Tidworth Military Hospital for an operation to remove possible residue left behind by the tracer bullet burning in my ribs. There was a lump in my side which was quite painful if pressed. They kept me in hospital about three days, during which I had a botched operation with local anaesthetic. Nothing was removed. The anaesthetic wore off at one point and had to be boosted. The surgeon went too deep for the anaesthetic a couple of times and knocked half the clamps off the blood vessels at least once. By the time he gave up he wasn't the only one who had had enough. They decided I would have to return at a later date to have the elusive junk removed under a general anaesthetic. I never did get it removed. There was no way they would have got me back there to be put under.

I thanked my lucky stars the Chinese surgeon who removed the bullet was not like that clown. My complete faith in British, and especially British army, doctors took a severe knock at Tidworth but, thank goodness, I have seen that bog-up redeemed many times since.

Another experience with the army medical scene soon after arriving home was also a bit off-putting. Less than a month after arriving home I was ordered to report to the British Military Hospital at Wheatley, near Oxford, for examination by a neuro-surgeon and a nerve specialist, to reassess the possibility of an operation to relieve the pain in my arm. Arriving at the appointed time, expecting to stay for a few days, I was taken to a ward and given a bed. It was late afternoon when I arrived and I didn't take a lot of notice of my surroundings.

Then I began to realise something was wrong. I had seen a couple of soldiers with rifles just outside the ward as I entered, but thought nothing of it. At first the other patients in the ward seemed the usual crowd one expects in such places but after a while I realised most of them were acting rather strangely. For instance, an oldish sergeant-major had approached me soon after I arrived and told me a joke – something about nuns and penguins – which he seemed to find funnier than I did. Five minutes later he was back, telling me the same joke again. It was after about the fourth or fifth time of hearing the same joke from the same man that I began to look at the other occupants of the ward more closely; I realised they were all mental cases.

The significance of what I then realised was an armed guard in the corridor suddenly tied in with other things I hadn't noticed before – like bars on the windows, heavy padlocks on the emergency fire-exit doors and absolutely nothing in the place which was heavy but throwable!

I was imprisoned again. A feeling of panic crept into my mind and became stronger as time passed. I wondered if the army doctors had found something wrong with my mentality and not told me. Was I expected to go mad at any moment? I thought back over the hectic recent weeks, of my release and journey home. Of the happy times during the last few days. I suddenly felt it would be good to hear that whistle blow – to be a prisoner of war was better than a prisoner in a mental hospital.

My mind raced but found no answers, no reasons, no calming facts. I approached an orderly and asked to see the sister. He said, 'She will be in in a minute, you can see her then.' When I asked again later, another orderly said, 'She has gone off duty now, you'll have to wait until tomorrow.'

What it obviously amounted to – nobody was interested. How could they be? Half the bloody ward wanted to see the sister, the doctor, the CO (the Queen?). The night did not bring a peaceful sleep, but by morning I had resolved to control my feelings, try to act normally, have faith in myself and my sanity.

I don't remember what I did or said but I must have got through to the sister, and when the matron did her rounds about mid-morning she said she would make some enquiries at once. In fact the matron used the phone in the sister's office, then came back and told me to bring my kit and leave the ward.

It had only been one of those little administration failures which bedevil even the best-run organisations. The fact that I had been booked in as a nerve case caused it. But it was one of the most terrifying times of my life.

The hospital authorities apologised to me for the mistake and sent me back on leave without seeing the specialists. I think I was on my way home within twenty minutes of leaving the ward, thanking my lucky stars for a sister and a matron who were people who believed in doing their job properly.

When I eventually returned to BMH Wheatley for specialists to see me, I was not detained overnight but was very careful to watch where I was being taken within the hospital – likewise on all subsequent visits! The specialists found there was nothing they could do to help the arm, but they seemed to think I was doing very well on my own.

About four months after my return to UK I had to report to Depot, The Gloucestershire Regiment, Robinswood Camp, Gloucester. Having reported in, I was told to take another month's leave. So back home I went for another month of rest and recuperation.

One thing which bothered me somewhat during those months on leave, and another reason I didn't like going into towns, was that several times I had a very, very strong feeling I was going to wake up in that camp again. The feeling was so acute one day that I had a stupid urge to jump under a bus which was passing, so the dream would end before I got to like it too much and the awakening became that much worse.

The pain in my arm was part of the problem. I began to reason with myself, if it was real our people would have stopped that pain by now. The many dreams of release I had in the camp had always been accompanied by pain, so I associated the two to prove one another. Too difficult to explain, but I had several stupid moments and they always happened in town. Luckily I found I already liked the dream too much to end it myself. So I resigned myself to wait for the whistle.

Upon my eventual return to duty at Depot Glosters my army medical grade was P7, the lowest I could be and still remain in the army. I had been offered my discharge on medical grounds but had asked to stay in, as I knew the army would do everything possible to repair the damage and get me fit again. The problem was that a soldier with only one arm is not a lot of use. The Glosters, however, put me into jobs I could cope with, while continuing to work on the arm, the pain and my general health.

Within twenty minutes of arriving at the camp I was in the nick! It was all rather comical, and it was seeing the funny side of things which put me inside.

Walking along a path towards the orderly room a rather high-pitched voice yelled, 'Hey, you!' Not being in the habit of answering to 'Hey, you!' I took no notice. The 'Hey, yous' got higher, louder and rapidly much nearer. Then it changed to: 'That soldier! That man!' Then (nearly at top C): 'Stand still!'

The clatter of fast-approaching ammo boots made me stop and turn. The man was just over five foot, but when he reached me, his jumping in the air was almost making us see eye to eye – only in the physical sense though. He was a sergeant or staff sergeant, well pressed and gleaming, with a big red sash to identify him as depot orderly sergeant. (A daily or weekly duty performed by all sergeants in turn.) We were about two feet apart and he was yelling so loud in his rather high-pitched voice that he couldn't get the words out properly. Eventually it got through to me. 'That is not a path!' Looking back at the rather well-trodden track behind me the thought occurred that even by army criteria this poor demented little thing had flipped off his rocker.

I probably showed my surprise, if not my pity, so he screamed, 'Read Depot Standing Orders – you will not walk on the grass!' The sense of humour was difficult to contain and likely showed a bit. I didn't get the next few screeches – too much volume – but then it came through loud and clear – pointing at the guardroom: 'Put yourself inside! At the double! Put yourself inside!'

Trotting to the guardroom, about one hundred yards away, I collapsed in laughter at the desk and couldn't even speak to the astonished Provost corporal. Five minutes later, having put the Provost staff in the picture, the clatter of fast-approaching boots of authority were heard and the diminutive DOS appeared in the doorway, a completely changed man, partly because he was now out of sight of the adjutant's office windows. (The adjutant at that time had a hate campaign on grass walkers.)

Within a very short time we had established a common friend – the little sergeant's big brother was on his way home from a Korean PoW camp. No charges were made and we were quite good friends from then on. Nevertheless I could still wind him up by saying, 'Mind that grass!'

My first job was in the Regimental police, which lasted a few months. To say my attitude towards the prisoners in the guardroom was too cushy would probably be an understatement. There was a thick early-morning fog and it was still quite dark at 6 a.m. when the prisoners were marched to the cookhouse for their breakfast. Knowing who was duty officer, and knowing there was no way he would be there that early in the morning it was safe enough to give the prisoners a bit of relaxation with American marching patter which, by then, they all knew pretty well. There were about a dozen prisoners, all shouting their parts at full blast:

'You had a good job and you left. You're right. Sound off.'

Then the bristling shape of the adjutant loomed out of the fog. He was standing in for the real duty officer. The good old laidback sense of humour as remembered so well from the battalion was totally absent in the depot. I'd had enough of that guardroom anyway!

Then to the Education Centre, from where the Education Officer sent me on a Unit Education Instructors course at 16 AEC Bovington. Having passed with high marks I returned to Gloucester and instructed for two or three months – until the army decided I had not had the qualifications to go on the course in the first place.

The next job was in the QM stores where I was QM Storeman for about two and a half years, by which time I had managed to get my medical grade up to FE again and volunteered for the SAS. The powers that be immediately had me posted to the Depot Training Wing as an instructor, where I spent my last two or three months in the Infantry.

In 1955 Ann and I were married and, afterwards, lived in an army hiring which was at the top of a farmhouse, at Longford on the outskirts of Gloucester. While working in the QM store in Gloucester, there were times when I had to work late at night preparing kit for a big intake of National Servicemen. On a few occasions, when Ann had nothing better to do, she came in to give me a hand. The main problem in a store like that is making sure other people don't 'liberate' anything. Ann just had to be different: I caught her slipping the odd bar of chocolate into a few kitbags. She claimed that she couldn't bear to think of all those poor lads being dragged away from their mothers. My comments about what it might cost us to compensate for about a hundred doting mums went unheeded – as usual. But I made sure she bought no more goodies to bolster the morale of the Army.

When I volunteered for the SAS I told Ann it was a unit which dropped supplies to troops in the jungle. She was probably a bit suspicious, as she had found out by then that I had volunteered for Korea.

Anyway, this was 1957, the SAS were on operations in Malaya, so we both knew a certain amount of separation was on the cards if I passed selection. As usual, Ann stood by me and we faced the future as a united family.

So ended my Infantry experience. Again I was a leaving many good friends behind. Apart from the odd clown, the Glosters were – and most likely still are – a damn good unit. I would like to think the air of laidback, confident efficiency which, to me, was the outstanding characteristic of the Gloucestershire Regiment in the Imjin battle, had somehow been carried on through the years to the present day.

As a young inexperienced soldier I was no doubt influenced by the attitude of those older, well-experienced men around me. Their unflappable example made me better able to cope, to do my bit, as did hundreds of other trogs in 29 Brigade.

Most of us were able to do our job to the best of our ability, but there were those who responded far beyond the call of duty and performed feats of courage and endurance under the most horrific conditions. Many of those are known only to a few amazed onlookers, but some were acknowledged by receiving a medal for their bravery.

The spirit of units like the Glosters can surely never die while they remain active units within the army. They will have their ups and downs but, in the end, come the crunch, the same spirit would again sustain men in the face of overwhelming odds, should the need arise.

Over the many years now, since the Korean War, people have often asked me if I think of Korea or remember anything about it.

As far as remembering goes, this book must speak for itself, I kept no diaries or records of any kind during my military service. The dates are correct from reference to various documents and factual histories, but more often from reference to my Record of Army Service.

Do I ever think of Korea or the Korean War? Yes, I think of the Korean War, just as thousands of others who have been left with constant reminders. However, I count myself extremely fortunate not only to have survived the experience but to have lived my life more or less as I've wished ever since. The memories, even so, are there if one digs deep enough: the sight, the sound, the smell, the feel all come back.

The Land of the Morning Calm was, to me, not all war, destruction and despair. It can be a very beautiful country, especially in the autumn when the hills and mountains with their cloak of trees display all the vivid colours of nature.

Perhaps my view of Korea in the autumn, which was, of course, from the confines of a prison camp, was also coloured by the fact that what I could see was outside the camp, therefore representing freedom, the outside world, untouched by the squalor, death and disease of my immediate surroundings. Nevertheless, it is a beautiful country.

7 SELECTION

Time. Time to think, Perhaps the best time to think, to take one's mind off the aching limbs, the searing pain of the back strap and the continual cutting of the shoulder straps. What does a man think of when he's flogging around an SAS Selection Course? Mostly: 'What the hell am I doing here?' But then, if you don't know the answer to that – there is no point in being there!

The forever hard going of the Brecon Beacons, and all points nasty for miles around them. The Black Mountains, Mynydd Eppynt, Fforest Fawr, away up past Painscastle (so aptly named). Everywhere was steep up or steep down, except where there were bogs or rivers. The rain only seemed to stop when it turned to sleet or snow on the high ground. The wind always made sure everyone got their fair share. If it slackened in its icy blast at night you knew you were about to walk into a tree or over a cliff.

Think hard and think about anything – except the discomfort. Not easy for most but, perhaps, a little easier for me. Pain and discomfort I knew well; they were no strangers.

So, here I was, after four years at Gloucester where for much of it I had been excused from just about everything bar breathing – a most unlikely candidate for the roughest, toughest selection and training in the army. But as any SAS man will tell you: it's all in the mind!

That's as may be, but it certainly doesn't feel like it when you are knackered out of your mind, facing an icy blast, aching, sore, soaking from head to toe and looking up another bloody great hill. Nevertheless, I felt I had the edge on most of my compatriots when it got really rough. I could always console myself that I didn't now have dysentery, a dangling useless arm or a yelling Chinese soldier threatening to replace the dysentery with a bayonet. There was always the thought, 'When it gets as rough as that, I'll think again!' Not a lot of use really, but then – it's all in the mind.

Ann had been a great help with a difficult rehabilitation, always keeping her sense of humour at the right times. How could I face her again if I couldn't pass a selection course? She always had faith I would recover, be fit again. This was the proof, so it had to be done.

There was also another little matter which helped with the get up and go to join the SAS. The Special Air Service was known to be a strictly operational unit, with no time for drill or bullshit. I had joined the army to be a soldier, had now served eleven years in the Infantry and had been drilled to bits and nearly drowned in bullshit. None of it had done me a blind bit of good when I had been in Korea. But the good training we could have done in the time wasted on the drill square would have helped a lot.

All I had to do to opt out of the stupidities of drill was to use my head and get my arse over the next hill. In 1957, the Selection Course for 22 Special Air Service Regiment was based at Dering Lines, Brecon. It was another of those wartime camps, a lot of wooden huts for the troggery, a few brick buildings for administration departments, all linked together by concrete roads and paths.

The SAS Selection Detachment, which ran the Selection Courses, consisted of one captain, one sergeant and three or four trogs. Two three-ton Bedford trucks and two quarter-ton Austin Champs were its wheels. Two or three Nissen huts for accommodation, another for office, stores and armoury made up the static assets. Meals were provided by a cookhouse serving other units based in Dering Lines. I found it a shoestring affair, casual but totally efficient. My first shock was seeing and feeling a loaded Bergen rucksack for the first time in my life, then realising I'd have to carry one.

There were four men on the course who had been given a second chance from previous courses due to injuries, illness, bereavement, compassionate leave or military requirements which caused them to leave the course before they could finish it. From them we got some idea of the problems ahead. With no small amount of surprise I recognised one of the troopers on the Selection Staff as a fellow ex-prisoner of war. Bert had joined the SAS almost straight from the PoW camp.

The course assembled on a Saturday. We were given a quick briefing and introduction by Captain Mike Jones, the Selection Officer. Then we were given a more detailed run-down of what was expected of us – and what was not expected of us – by the sergeant. He was Paddy Nugent, one of the greatest men I've ever known. He was certainly the best instructor I ever met in the army – as Irish as the Blarney Stone, with quick wit, a great sense of humour and a turn of speech only believed when heard.

On the Sunday we prepared our kit, then were given lessons in map reading and navigation, with great emphasis on the use of the compass. I had been involved with instructing map reading, both in the Education Centre and with the Continuation Training Platoon at Gloucester. I had also taught myself a lot about navigation at various times, but, although the lessons on that Sunday were voluntary, supposedly for those without much idea – and there were many – I joined in. There is always more to learn in such subjects. I learned a lot! Part of the Sunday lessons involved a walk into the Brecon Beacons, which proved invaluable for getting switched on to the terrain for the coming weeks.

My second shock came early on Monday morning. I had been over some rough old assault courses in my time, so, when they said we were going on the assault course I thought of the worst possible case and was ready for it. When I saw the assault course I thought they were joking. It was a doddle: One six-foot wall, a nine-foot jump over some water, a few bars and ropes – a piece of cake.

And so it was a piece of cake, for the first ten times around perhaps, then I began to notice it a bit. I've no idea how many times we went round that snotty little course, Paddy Nugent lashing us with his tongue and making dire threats to lash us with other things too!

At the end I was fit to drop, completely smashed. Some had already dropped. We were told that was just a warm-up. Sure enough it was. My next trick, as far as I can remember, was a thirty-five-mile hike, with fifty pounds on my back, all across country ('you will not use roads or bridges') which lasted nearly two days. We went from one RV (rendezvous) to another, never knowing which would be the last. I waded the River Wye twice and slept the night on the mountains, soaking wet and cold but too shattered to notice. We had to travel alone, if two men were caught walking together they risked RTU (Return to Unit); the same punishment if we were caught on roads or bridges.

As the days went by the numbers on the course dwindled. Some men gave up and asked to be sent back to their units, others were sent back because they couldn't keep up, one or two were injured.

At last, the final exercise started. It was the endurance march, which lasted two days. The end of the first day I had a problem: I had pulled the Achilles tendon in my left ankle; walking became difficult. The last RV of that day was in the mountains and we all had to sleep at the same place. On the last half mile to the RV I was on a grass track when a Champ came along with three of the staff in it. They tried to talk me into getting a lift to the RV. Then they said I should pack up altogether. I walked on. There was no way I was going to pack up on the last bloody few miles.

The weather was so bad we were given a rum ration that night.

I saved mine – plus a few from others who didn't like the stuff – so that in the morning I could deaden the pain in my ankle.

Morning came, I deadened the pain and ploughed on. The first RV was at the Storey Arms (not a pub but a café then) where we had a weapon inspection. The last RV was at Talybont Reservoir, which meant crossing the Brecon Beacons again, a distance of about seven miles.

The ankle slowed me down and I knew that they wouldn't wait for me. I was about a mile from the RV when I saw the truck pull out, heading back to camp. I checked my watch and made a note of the time. Once down to the road I got a lift in a car to Brecon and hobbled into camp to report. At the office I told them where I was when the truck pulled out. They said, 'Prove it.' So I told them the time the truck pulled out; I could see I had proved it.

Half an hour later we were called to the office, one at a time, to be given our results. There had been twenty-seven men on the course when it started. Six of us passed – the four who were given a second chance from previous courses, myself and one other. The six who passed were given the weekend off and told to report back for training at 8 a.m. on Monday. The others were given rail warrants to get back to their units as soon as possible.

I felt great at that moment and couldn't wait to get back to Gloucester to tell Ann I had passed. Without bothering to shower, I put on my shoes as the boots were too painful, jumped on my bike and headed for home.

Just before I reached Crickhowell a coach coming the other way turned across the road in front of me. With nowhere to go I had to hit it. Back to the familiar sounds and smells of hospital, this time as Newport's Royal Gwent Hospital, with a badly injured left foot. The foot needed thirty-odd stitches and was put in plaster for three weeks. If I had been wearing my heavy army boots instead of light shoes it's likely I would have got away with bruising.

I stayed in hospital for nearly a month, and found the difference between the SAS and other units I had been with. Never, in all the time I had spent in hospital, had I ever been visited by a representative of my unit. I had seen officers, padres, sergeants and sergeant-majors visiting other men many times but I had never seen a representative of my unit while in hospital.

Though I could hardly claim to be a member of the SAS, having only passed the first part of the entrance exam, Captain Jones came to see me the day after I went in. Paddy Nugent, Bert and the others dropped in every few days, sometimes at very odd hours and in very muddy field kit. Thanks to the staff at the Royal Gwent they were allowed to see me any time of the day or night.

Ann travelled down from Gloucester as often as she could, so altogether I had all the help I could get. On one of his visits, Captain Jones said the Glosters were honking because they had been unable to make the final transfer, and didn't know what to do about me. He said not to worry, he would hold the fort until I could find out how the foot would shape up.

The surgeons were pleased with progress when the plaster came off. Soon afterwards I was allowed to go home, provided I report to the Army Medical Centre at Gloucester every two or three days to have the dressing changed and the wound checked by the Medical Officer. As soon as possible I hitched a lift to Brecon to pick up my motorcycle.

Once I had the bike again it gave me the mobility to get over to my folks' place in the Cotswolds, a lonely cottage among the hills and woods a couple of miles from a village called Guiting Power, about twenty miles from Gloucester. It had been our home since 1939. I knew every inch of the surrounding countryside which was now my training ground in the struggle to achieve physical fitness. Every chance I got was spent hobbling, then walking, then running around the hills and through the woods.

Sometimes I carried a shotgun and played my favourite sport. Walking through a fir plantation, between the rows of thirty-foot conifers to disturb wood pigeons, which would take off with a clatter of wings. If they passed over the narrow gap above me, nine out of ten were dead. This called for split-second reactions and was to stand me in good stead for the future. It was probably the best training anyone could get for instinctive shooting.

My parents supported my efforts in every way they could, as always, although my mother worried I would do more harm than good when, in the early days, she accidentally saw a blood-soaked bandage. Ann and my parents between them made the task a lot easier than it might have been, but then, they had been through it all before, when I returned from Korea. The injuries were different, but at least the mind was more settled this time. Ann's parents always made us welcome too. A lot of time was spent at their place, near Tetbury, where we hoofed around the lanes and roads every time we could.

All went well but it was another five or six weeks before the swelling went down enough to get any army boots on – even then I had to cheat a bit. Thanks to having friends in the QM Department at Gloucester, I managed to get a pair of boots two sizes too big, so I could use a bigger boot on my left foot. Captain Jones told me on the phone there was another selection course starting in two weeks. If I thought the foot was okay to try out, I could go down to Brecon and give it a go.

So, back to Brecon with my odd boots – which I didn't mention to the SAS staff, of course! Captain Jones asked how the foot was. I told him it was okay, so he said to try it out on the first march of the course. This I did and found it no problem, although the wound was still not properly mended and I had a bandage on it. Captain Jones asked me how it was after the first march. When I said it felt all right, he said, 'Good, give it another try tomorrow.' And so on, until I had completed another selection course, and passed again. I 'wuz conned', but there can't have been many who had to pass the selection course twice to get in once. (I had to retake the course as it was only an entrance exam and the circumstances had changed with my injury; I had to prove that I was still strong enough to join the SAS).

The old shoulder problem had given me a hard time once in a while, but all those years of tedious exercise and continued practice of blotting pain from my mind had paid off. I had cleared the first hurdle and would take some stopping from now on.

A week or two of training followed, then it was off to Holding and Drafting Company at Airborne Forces Depot, Aldershot. The bullshit and the attitude of the snivelling para NCOs in 'H' and 'D' Company came nearer to cracking me than anything on the selection course. Luckily for one or two at Aldershot, we had too much to lose to start throwing punches. They probably knew that as well as we did.

Away from the Glosters very few people, even in the army, recognised the ribbon of the United States Presidential Unit Citation which was awarded to the Glosters and the men of the battalion who were in the Imjin River Battle. It was worn on the sleeve, just below the unit title flash, and in Airborne Forces Depot, where almost everyone had para wings under their unit title flash, the citation drew many curious glances. Standing in the queue for char and wads (tea and sandwiches) in the NAAFI one day, a group of young paras, with brand new wings, asked me what the ribbon was for. Before I could answer, one of my friends (an ex-para) said, 'That's the Blue Coffin. Haven't you seen one of those before?'

'No, what's it for?'

'It's for two thousand jumps of course. What else?'

Over the years the citation has been explained away on many occasions. Another good one I remember was, 'Oh, he used to be a "Lootenant" in the Seventh Cavalry!' That one to some US Marines.

After a week or two's embarkation leave we flew out to Singapore, then by train to Kuala Lumpur, capital of Malaya, and base for 22 Special Air Service Regiment. Malaya, once part of the British Empire, had been working towards independence since the Second World War. The Federation of Malay States became an Independent State and elected to remain within the Commonwealth in 1957.

Japanese occupation in the Second World War, during which anyone who would fight against the Japanese had been armed and trained by the

British, saw the emergence of the Malayan Communist Party as an effective fighting force. Its members were mostly Chinese, and less than three years after the end of the war, Mao Tse Tung's revolution in China may possibly have given them the idea of taking over Malaya and making it a communist state. As at least seven-tenths of Malaya was covered by (so-called) impenetrable jungle, the terrain was ideal for terrorist operations. In 1948, the communists tried to take over.

The Malayan Emergency lasted about ten years. During the early days the Brits had a pretty thin time and cock-a-hoop communists paraded through Ipoh, the northern capital of Malaya, with flags waving and bands playing.

In my humble opinion there were two very much overlooked factors that prevented the Malayan Emergency from becoming another communist takeover. Firstly, was the fact that the people of Malaya didn't hate the British or the British Army, an army which does not alienate itself from civil populations. Secondly, was the army's old-fashioned close-and-destroy methods, which put men on the ground instead of a preponderance of high explosives and fire-power.

Ninety per cent of winning or losing is a matter of morale. A man who has been chased off a position by a few trogs with rifles and bayonets is going to feel a damn sight more beaten than one who has sensibly withdrawn from an area being blown to bits.

The communist terrorists (CTs) turned every road and railway in Malaya into a potential death-trap by continual ambushing. But the trogs were undeterred and often turned an ambush into a death trap for the ambushers. Slowly, at first, the army gained the upper hand. Deep penetration jungle operations, part of the close-and-destroy policy, were required, and (more by accident then original intent) the SAS were reformed in 1950 to do the job.

By the time I arrived in Malaya, the CT was a pretty rare bird. Someone claimed the comparison was like a grouse shoot in Yorkshire – with only ten grouse in the county. Nevertheless, these particular 'grouse' were very good at shooting back, or even shooting first.

I think it is worth mention that the Malayan Emergency was brought to a successful conclusion as a result of the Commonwealth effort. British, Gurkha and Malay trogs took the brunt of the fighting but some damn good troops from other Commonwealth countries also made their mark on the scene. The British Commonwealth is one of the very few organisations in the world to have beaten a well-organised military coup by the communists, and in a country with terrain which greatly favoured the opposition and which made the machinery of modern warfare less effective than it was in Vietnam.

When my small group of trainees arrived at Wardiburn Camp, just outside Kuala Lumpur, it was November 1957, and we were amazed to find one of the squadrons having their Christmas dinner. Then we found they were going into the jungle on operations for four months. This was a whole new world, there was a whole new language with it too. Not only all the old army words, picked up in India, Egypt and many other places, to be forever mispronounced, but now a heavy dose of Malay. We were baffled by 'buckets', 'bashas', 'Kualas', 'rotan', 'attap', 'muckan' and a host of other words used in everyday conversation.

Our next move should have been a training operation but, for some reason (I never found out what), we were sent on a parachute course. The para course was run by RAF Changi, Singapore, so back down to Singapore we went, after only a few days at KL, to start something I had always wanted to do, but had some doubts in my mind as, although I didn't know it then, I have vertigo – a fear of heights.

There were two things I hated on the para course. One was the fan. It was touch and go whether the fan itself or the climb up the rickety steel ladder to the platform was worse. Our introduction to the fan was a typical Para School con. One of the instructors, called 'Titch' (for obvious reasons), gave the demonstration. He was maybe seven stones wringing wet, and when he came down he floated like a bloody fairy. I thought it was no problem and was first up the ladder. Putting on the harness, I didn't much like the look of the thirty- or forty-foot drop to the matting, but I thought of how Titch had sailed down, so I threw my sixteen stone into space. I came down like a bomb and was amazed to be able to stand up afterwards, let alone walk away from it. Later I found it was wiser not to throw my sort of weight out so far off the platform. That way the fan could exert a bit of brake power!

The other contraption I detested was the 'bikini'. This was a harness used with a long tape, to get down from jungle treetops after parachuting into them. (An exclusively SAS method of entry into the jungle.) We were trained in the use of the bikini by lowering ourselves from a platform about thirty feet high. The bikini was a very cunningly devised instrument of torture. It was like a heavy canvas pair of swimming-trunks, hence its name, but it had a steel ring at the front, about six inches in diameter, through which passed the tapes supporting your weight.

We had to do the training with a heavy pack, as per operations, and the torture was partly due to the pack. We started by hanging in a parachute harness from the platform. The next move was to tie the tape to a secure point, then put your weight on to that by releasing the parachute harness. This move made the top of the steel ring try to push through your stomach to your back-bone. Painful but bearable. The next move was to drop the heavy pack. This caused the ring to change its direction. It now stabbed down at your wedding tackle and did its best to destroy it at every move. Very, very painful. In my case, not bearable for long. I dreaded ever having to parachute into trees with that thing.

There were no balloon jumps in the Far East parachute course, we did all our jumps from a twin engine Valetta. My first jump was great. I really enjoyed it – until the ground hit me! Again my weight had something to do with it. Being so tall had its disadvantages too.

The British military parachute of the day, called the 'X Type', was very good. It was designed to get the average soldier from the aircraft to the ground as quickly as possible with a minimal chance of injury – as long as he did everything right. A slower descent could, given the uncertainties of local wind conditions, cause problems with the accuracy of the drop and could result in a wide scattering of troops on the ground. Also, hanging around in the sky is not where one wants to be if there's any chance of being shot at. The problem for some of us was being a bit too much above average weight – in my case about sixty pounds above the average – which gave us a much faster descent. We always claimed the 'PX' parachute, which came on the scene a few years later, actually slowed us down, whereas the old 'X Type' only kept us upright!

My first jump with equipment brought out one of the problems of being tall. Our first three jumps were 'clean fatigue' (no kit) but for the fourth jump we had a container for equipment, which weighed about forty pounds. For the jump from the aircraft the container was hung from two hooks on the front of the parachute harness and held lower down by a leg strap around the right leg. The bulk of the container was from the waist to below knee level. Before landing the container was released to hang below us on a fourteen-foot nylon cord.

The exit door on the Valetta was not too high, causing me to duck as I went out. This ducking combined with the extra weight low down induced a beautiful forward somersault outside the aircraft, resulting in me wrapping myself in the parachute as it developed. Realising something was wrong, I opened my eyes and looked up. (I never did a 'daylight' jump after the first two!) Where my chute should have been, all I had time to notice was the big red cross on the roof of the ambulance attending the drop. I looked down and there was what appeared to be a big bundle of dirty washing wrapped around my feet.

As usually happens in a crisis, everything slowed down so I had time to think. I had been well trained, so I went through the routine. Made a quick attempt to free the main chute. When that got me nothing I pulled the reserve chute handle and started to deploy the reserve canopy. I'd only just started on that when the main chute partly freed itself, changing the situation. The words of my instructor, 'Never deploy your reserve if the main canopy can be used,' came to my aid. I gave the main canopy a bit of help and it developed, slowing down my rate of descent quite well. I was tangled in the rigging lines and still upside down, but now I had time to work on that. I undid my container, and pushed it up through my legs. Its weight then pulled me horizontal. I was now on my back and could better see what to swear at.

Glancing down I could see there wasn't much time. People were running towards me with no hope of making it. Eventually only one foot was still caught in one line so I kicked the line off with the other foot. As my feet swung down they hit the ground. Good exit, lousy flight, but the best landing I'd had so far.

One funny thing about that incident was that there was an instructor on the DZ (Drop Zone) who had a loudhailer to talk trainees down and tell them if they do anything wrong and what to do to put it right. We were dropping two at a time that day and I was number two (second out). I'm told the loudhailer patter went like this, 'Okay, number one, your canopy has developed – look up! – Okay, number two, your canopy has . . . F***ING HELL!' We heard there were a mass of complaints from the RAF married quarters which adjoined the DZ. Those loudhailers were really loud.

On all para courses at Changi a teddy bear in SAS uniform and with its own little parachute was thrown out every time the aircraft made its first drop run of the day. This was the course mascot. At the end of the course it was raffled to raise cash for a local orphanage for the disabled. The afternoon before the course finished we all went to the orphanage and took the kids to the beach for a few hours. On my particular course the teddy bear was put back in to be raffled three times and eventually given to the orphanage. We also took them a pile of sweets and various goodies. They enjoyed it and we had a good afternoon in the sea which we enjoyed too.

After successfully completing the para course we were allowed to wear a khaki parachute with no wings on our lower sleeve. As our sleeves were always rolled up in that climate I doubt if anyone ever sewed them on. We had a long way to go before we would be allowed to wear the wings of the SAS. To qualify for SAS parachute wings we had yet to pass the test of a training operation in the jungle. This involved at least a month in the jungle, being trained in basic SAS jungle patrolling.

Before going on the Training Op we were taken into the jungle near KL, to be taught the rudiments of jungle movement, how to use our equipment to make shelters (bashas) and hammocks, how to make the most of our rations and general jungle behaviour. For this introduction to the jungle we were only 'in' for two days and one night. 'In' always meant in the jungle. 'Out' always meant out of the jungle.

It was on this first visit to the jungle that I met what was to be one of my greatest enemies: attap, a useless, savage plant which grows in profusion in the jungles of Malaya. It has a clump of fronds rising from ground level, splaying out in all directions, up to twenty feet long. These have nasty thorns on the underside of the stem. But the real wicked part is like a long bullwhip. Some plants have several and they can be very long, probably up to forty or fifty feet. The plant holds this whip like a fishing rod. It can be two or three inches thick at the butt end, tapering down to almost a hair at the lash end. The lash end can be dangling to touch the ground several yards from the base of the plant. Mixed among the other greenery of the jungle it is often almost invisible. Every six inches at the butt end down to every half inch at the lash end there are very sharp hooked thorns – in fours, around the whip – the whole of its length. It is very strong, not possible for a man to break by pulling, except perhaps at the very fine end, maybe the last couple of inches. Getting away from attap, once caught, is a delicate operation which takes practice – and patience. I never could see the point of catching anything which moves just for the hell of it.

8 MALAYA

A few days after our introduction to the jungle, we checked our kit for the last time, drew rations, maps, compasses, weapons and ammunition and climbed aboard the trucks which would take us to the area we were to use for the Training Op. I don't remember how far we went by road, but I think not more than twenty miles. Then on to our flat feet. The order came to 'saddle up', we heaved our Bergens on to our backs and began the slog into the unknown.

We were carrying two weeks' rations which, by SAS standards, was not a lot. Twenty-eight days' rations was the norm when moving in. Our weapons were the standard SAS weapon of that time, the Belgian FN rifle, forerunner of the British SLR (Self-Loading Rifle). Each of us carried five twenty-round magazines. The main difference between the FN and the SLR was that the FN had a three position change lever giving automatic and semi-automatic fire, whereas the SLR can only fire semi-automatic. We found the FN was too inaccurate when fired on automatic so the SLR was really just as good. The other main difference was the flash hider which, on the FN, was incorporated in the bayonet. The SLR had its flash hider as an extension of the barrel.

My Bergen probably weighed seventy pounds (I'm a big eater!) and by the time we stopped that first night I knew about every ounce of it. Jungle-marching is nothing like any other marching. We were forever climbing over fallen trees, or crawling under them, getting tangled in attap, sliding in mud, tripping over an assortment of roots, vines, sticks and stumps. There was nothing like it in the *Tarzan* films, or any other damned film I'd seen. Having been brought up in the woods I felt quite at home in the trees and had less trouble coping than most. Nevertheless it was very hard going.

At last, after, I think, two days' march, we reached a place which had been selected to set up a base camp. Next morning we set out on a light-order patrol for the day – that meant leaving our Bergens in the base camp. The first snag was crossing a raging torrent not far from the base camp. There was a line across the river, but it was made of rotan, a type of vine used as ropes by the jungle people. (Not the *Tarzan* type.) I was quite used to using vines back home in the UK, but not to take my weight over anything dangerous, as the only vines I had ever used were extremely unreliable.

When my turn came to put my faith in that rotan, I approached the prospect with some misgivings – actually, that's understating the case; I was bloody terrified. The river was, as mentioned, a raging torrent. The water was deep and brown – except where it was churned up into creamy spray by protruding jagged rocks. It was about forty yards wide at the crossing point, but much wider about one hundred yards downstream, where another equally nasty torrent joined it. The only consolation was that nobody would hear me scream if that bloody rotan broke – the roar of the two rivers made sure of that. My extra weight made it sag more than it did with the others. The water did its best to tear me loose, but it didn't break. Nevertheless I wasn't looking forward to the return trip that night.

After a couple of weeks patrolling from that base camp we cleared a DZ for a resupply air-drop, which came right on the button. Then, loaded down with two weeks' rations, we marched again for a couple of days to make another base camp.

For the move to the second base camp the twenty-eight men on test were divided into six or seven patrols, each moving by a different route, accompanied or led by a staff instructor. One patrol failed to arrive at the new base and, after a wait of twenty-four hours, search parties were sent out. I was with the patrol which found the missing men. The instructor with the missing patrol had two bullets in his thigh (I never heard how they got there), so the immediate problem was to get him evacuated from the jungle.

The instructors decided the best move would be to get the injured man to the base camp, which was about three thousand yards away, as the manpower was there to clear an LZ (Landing Zone) for a helicopter.

With the extra help of the patrol I was with, we made fairly good time towards the base camp, but it became obvious we were not going to make it before dark. Two of us were elected to try to get back to the base camp so things could be organised by radio and the men could start work on the LZ at first light next morning.

We were then about two thousand yards from the base camp and had about half an hour or so of daylight left. The terrain was very rugged and the jungle, especially attap, was very dense. When darkness fell we were about four or five hundred yards from the base camp. We had no torches, being in light order, so headed for a small river which came down near the camp. Once on the river we were able to make use of the last scraps of light to scramble up waterfalls and wade through pools to make another couple of hundred yards in the right direction. Wanting to be sure in which direction to go, we left the river and moved slowly through the undergrowth until we judged we were one hundred yards east of the river. Then I knew the base camp had to be to the south of us, hopefully within two hundred yards.

The blackness was total. The attap was thick. The air was blue. But we slowly made ground until we could find no way through at all. Absolute solid walls of brush and attap every time we tried to go south, and I mean solid. We sat down, and I lit a cigarette. Time to think, I wasn't all that experienced in the jungle, but somehow I knew solid walls don't happen. There's always a way through.

Then it clicked. When we left that morning some of the men were going to start cutting a DZ for a supply drop. We had come up against the fallen trees, or at least the tops of them. Now I knew where I was. (The other man had been with the missing patrol so had never seen this area.) We moved further east and, at last, the way south was comparatively clear. Within ten minutes we were in the base camp – much to the surprise of the staff, who never expected anyone nearly two hours after dark.

Radio messages were sent that night, and first thing in the morning we started enlarging the DZ into a LZ. To this end, a light aircraft came over and dropped a 'packet easy' – a one-hundred-pound pack containing explosives, fuses and detonators – with which to blow trees and make a quick job of the LZ. There was one small problem, the drop missed the DZ by nearly five hundred yards and was nearly lost. It took another explosives drop to secure the first one, which was at the top of, probably, the highest tree in Malaya.

The normal method of recovering 'unreachable' parachute supplies was to shoot them down with a signal pistol, which set fire to the parachute. With a packet easy, however, it is essential to be well clear when it drops, and of course fire is not generally a good mixer with detonators and explosives.

By mid-afternoon the casevac (casualty evacuation) was complete. We got back to the hard grind of real training. One of the men on the training op was an old SAS soldier who had been home on inter-tour leave, so had to do the training op again to qualify. He was from Liverpool, so was called 'Scouse'. He was very fond of deriding my anti-leech precautions, taking great pride in showing how he didn't give a damn for leeches. I had found long ago that the best armour against infection is the good old skin. So I had gone off all things which make holes in it. Leeches seemed to think of nothing else but making holes in skin. Being specially designed for the job they are very good at getting through almost anything in order to get a good fill of blood. So I took my precautions – in this case applying mosquito repellent on my trouser legs and boot tops – and remained not too badly affected. Scouse, on the other hand, didn't even wear underpants.

We had been on a long hard patrol all day and were returning to base camp down a river track. The leeches were almost like blades of grass waving in the track, trying to catch a passing foot. We stopped for a break and Scouse undid his fly to relieve himself. Then he yelled and we all gathered round. He had eleven leeches on 'it'. It looked like a bunch of grapes hanging out of his slacks. When he pulled them off he bled like a stuck pig.

It got funnier. He had to wrap a shell dressing around it to stop the bleeding, then tied the tapes up round his neck. We did the last couple of thousand yards back to camp with Scouse waddling, legs apart, threatening to shoot us all for seeing the funny side. I've seen some things in slings, but that was the ultimate sling job.

The end of the training op came, and we had to march out in patrols. They made the mistake of putting me in charge of one patrol. There were four of us, we were all starving. We were all big eaters and had a job to exist on the rations we could carry. We had not yet learnt to pack the best things and leave out the rubbish.

We knew that at the first base camp a food cache had been left by the instructors, presumably to boost their rations on the next training op. The march out to the road head was a two-day slog. I worked out the angles and reckoned if we moved fast and were prepared to add about four thousand yards to our distance, we could make pigs of ourselves on the first night. I put the plan to the other three. The unanimous decision – we would eat! So we did. The tins of chicken supreme I yoffled that night were the best I'd ever had. We had a great feast and marched out the next day feeling full at last.

So ended a hard but very interesting experience. I think about twenty of us passed that test. I, for one, realised just how little we had learnt of the vast volume of jungle training. We had been taught a lot by excellent instructors, but it would take a lot more time and sweat before we were really good in the jungle.

Somewhere at about this stage in the game I lost my corporal's stripes and – as the SAS put it – was 'promoted' to trooper. In other words, I was starting where all real SAS other ranks start, at the bottom. Anyone who never got promoted to trooper can hardly claim to be truly SAS. The difference in pay was partly made up by para pay which was then two pounds and ten pence per week. I always claimed para pay was reasonable for jumping – but landing was definitely worth another fiver.

After the training op we were posted to operational squadrons. I went to 'D' Squadron's 16 Troop. Before the squadron went on ops I had time to nip down to Singapore to meet Ann off a plane from Britain. Ann and the wife of one of my best friends were installed in a bungalow at Petaling Jaya, a few miles from KL. Then we had a few days in which to get to know KL before the squadron went in. 'In' this time was the Triang swamps. It was a three-month operation.

Our method of entry was from a briefly stopped train at night. Not being keen on swamps I was glad to find this one was dried up for the moment, the main problem being to find any water for drinking and cooking. We pounded the area day after day in three- and four-man patrols, finding practically nothing in the way of enemy signs. A few old camps, long since vacated, ancient tracks, a couple of water points nearly rotted away, nothing less then several months old.

On one occasion we returned to our base camp after a seven-day patrol and when I opened my Bergen – which had been left at base – I found a black shiny scorpion sitting just under the lid. It looked so much like a toy I thought someone was sending me up and went to take it out. Luckily my basha-mate, an old hand, saw it too and knocked my hand away just in time. It was very real all right. I still had a lot to learn.

During the last month of the operation three people went down with leptospirosis, a fever carried by rats. One was an officer, 'Biffo', one was Nat, an Irish trooper, and the other was an Iban tracker. They were evacuated in a struggling ancient Sycamore chopper, which couldn't get off the ground with three of them and had to leave one behind and come back for him later.

On the same op, one of 16 Troop corporals was marching with a temperature of 105. There were two corporals in the troop and one of them was troop commander, I was never sure which. We also had a lance-corporal in the troop. Ranks were not high or plentiful in those days. No one worried about rank, the only thing of importance was the ability to do the job.

The corporal with the temperature said it was just malaria, he would sweat it out. So he did, and carried on with his job. I was duly impressed, but didn't fancy trying it myself if the time came.

We had four aborigine porters with the troop who carried the spare radio batteries. Many aborigines worked with the SAS; they were the original inhabitants of the Malayan peninsular and almost all of them still lived in the jungle. They, and the troop signaller, stayed in the base camp while the remainder went out in threes and fours on patrols which were sometimes one or two days, sometimes one or two weeks. The patrols had no radios, the only contact with the outside world was through the troop signaller.

There were four of us in the patrol I was with, led by a trooper. We did our job thoroughly, no matter how hard the going, we never gave up. There was no movement at night. Just before dark we would stop and put up our bashas, two men to each. While we were fixing the basha, a mess-tin of rice would be boiling for the evening meal. Once it was dark we lit candles and read books to pass the time. In deep jungle there is rarely enough draught at ground level to flicker the flame of a candle.

A resupply by air came every twenty-eight days. We would all come into the troop base the day before it was due in order to clear a DZ. Resupply day was often referred to as when 'The *Mukan* Bird Sings'. *Mukan* being Malay for food. Wives were allowed to send parcels to their husbands. These were collected from our homes by the regiment and came with the air-drop. Ann knew my favourite food and always sent plenty of it. She also sent books and magazines. There were only two married men in 16 Troop, but the amount of food and things our wives sent was enough to share with the whole troop. On one occasion our two parcels had a parachute all to themselves.

The troop was a mobile library. We all went in with at least two books which were passed around. More came on the air-drops. If we ever came into contact with another troop there was always a big book-swapping session. A selection of newspapers, magazines and books were always sent on air-drops by one of the unsung heroines of the WVS (Women's Voluntary Service) who organised a library, games-room and reading-room at our base at KL. Though very seldom heard of, the WVS did a lot of good work behind the scenes, being much appreciated by the troops on operations.

We always got one day's fresh rations on the air-drop. Usually steak, potatoes, vegetables and fresh fruit. We also got new boots and new OG (Olive Green) shirt and slacks. Other items of military clothing and equipment were also sent on request. 16 Troop, while I was with it, always made the most of the air-drop, having a beano of food on these days.

The next morning, at first light, we moved base camp to a fresh area, often ten thousand yards or more distant. That which could not be consumed or carried was destroyed by burning. Anything which wouldn't burn was buried.

Some troops moved base camp immediately after taking their air-drop, which meant a lot of good food was wasted, as a man can only carry so much. I didn't think our one-day-per-month 'rest' was overdoing it. The operational situation at the time didn't put us on the defensive, or there may have been good reason for moving after a drop.

After the Triang operation, we had about ten days in KL, then came a week of relaxation and training at a beach near Port Dickson, on the west coast. The snag was that we had to parachute on to a beach near there.

It was the briefing for the jump which caused all the problems. They said it was a small beach with an area of jagged coral at one end and a mangrove swamp at the other. We were jumping from a Hastings and I was near the end of the port stick (left side), which made me worry about landing in the coral. There was little or no chat about the problem, but it was very apparent afterwards that everyone in the squadron was thinking along the same lines.

Standing in the aircraft as we approached the drop I noticed both sticks were getting more and more bunched as we at the back crept forward. When the green for go came on, there was a pause at the front end as they made sure of missing the mangrove swamp. Meanwhile the back of the stick was pushing hard to get out before the coral got under us.

The result was obvious, we came out of the aircraft on one another's backs in a big, fast-moving 'bus queue'. How I got out the door I'll never know. The RAF dispatchers were on their backsides beyond the doors and I was running before I got near the exit. There were no real injuries, but a lot of strop burns from contact with other people's deploying parachutes, also a lot of bruises and a few cuts. No one was anywhere near the mangrove swamp or the coral, the whole lot of us were in a big heap in the middle of the beach.

There was, however, a problem. The Commanding Officer of 22 SAS had not been with the unit very long. He had come to watch the drop, the first he had seen involving SAS troops. He was not impressed. Coming from the Parachute Brigade he was well used to seeing good, clean, regulation jumping. This 'absolute shambles' he had just seen was obviously the way the SAS always did their parachuting and it was not good enough!

A really horrific bout of para training was going to be the answer. In the event it didn't come off – thank goodness. Hardly anyone in the SAS liked parachuting, we looked upon it as a necessary bind; another means of getting from A to B, no more, no less. One of the things I noticed about the regiment, no one was ever afraid to say they didn't like certain aspects of their military situation.

After Port Dickson, the squadron went in again, I can't remember where, but for two or three months, maybe four. Again, although it was supposed to be a 'hot' area, we found nothing of note. A lot of sweat, a lot of hard patrolling produced the usual big fat zero. The opposition were getting very scarce by mid-1958. The only things of note I can remember of that op was one of my friends getting hit so bad by hornets that we had to stay put for a day to give the swelling time to go down so he could open his eyes. Another very good friend of mine had an argument with a phosphorous grenade, which we used as smoke grenades to help the Air Force find our little hole in the jungle.

On this occasion Don threw the grenade but it hit a branch and was catapulted back at him. With E-Type acceleration he nearly got clear but a dollop of phosphorus landed on his head and set fire to his hair. He dived in the nearby river and scrubbed his head until it stopped smoking, then came out of the water to where we were all laughing at him. A minute or two later someone noticed Don's hair was smoking again, so back into the river he went for another scrubbing. The same thing happened several times during the following half hour. I wouldn't mind betting if anyone looked with horror above Don's head that he'd get a very worried look in his eye.

On the night of the tigers, as I always remember it, the troop was moving base camp from one area to another and, for some reason, we stopped one night in a saddle between two high ridges. We had seen tiger tracks and I had caught a glimpse of one once. I only knew I'd seen a tiger when we found its fresh tracks where I had seen the flash of golden movement.

That night on the saddle I'll never forget. The tigers started roaring about midnight. We had all been asleep, but the first shattering roar from about one hundred yards away had everyone wide awake. That was answered by other roars and growls from all around. At the time I didn't know that tigers hunted in packs. For some reason I never thought of them attacking us, even when they were only a few yards away. As always in the jungle at night, it was inky black. I could hear and smell tigers, but couldn't even see my hand in front of my face. I felt distinctly nervous but not particularly frightened.

We had an Iban tracker with the troop (Ibans are from one of the head-hunting tribes of Borneo where – I believe– there are no tigers); he was so terrified he left his basha and sought human companionship. That nearly cost him his life as his groping hands landed on one of my friends who was lying there cuddling a shotgun. Their combined yells made me think at first someone was being taken for supper, but there was nothing anyone could do. We were all completely blind in that darkness. Our bashas were about twenty yards apart, many of us sleeping alone in one-man bashas – as I was at that time – so all we could do was wait. My greatest fear that night was not from the tigers, but that someone would start shooting. Eventually the big cats moved off, their growls and roars faded down the valley. All was well, sleep came quickly again.

The SAS operated in Malayan tiger country for about eight years.

Mostly in small patrols of three or four, sometimes only two men, and obviously one man moving on his own, sometimes for many hours. Yet I have never heard of any SAS man being attacked by a tiger.

On one operation I managed to frighten myself quite well one night. My basha was about twenty yards from any others, and I was on my own, reading by candlelight. We must have just had an air-drop as I had some empty cans in a box just outside the basha. Hearing the cans rattle, I turned to see what caused it, and just caught sight of something moving quickly into the darkness.

Grabbing my torch I shone the beam in the direction the animal had gone. A pair of large feline eyes shone from a few yards away. They belonged to something with a furry, black-spotted body, which looked quite big through the network of leaves and twigs. I immediately thought 'leopard', and for no reason I can think of I made that squeaking noise which is used for calling your pet cat. The animal immediately came towards me and in my rush to grab either my *parang* or rifle I dropped the torch and knocked over the candle.

Then I sat there in inky black silence holding the *parang* or rifle, I can't remember which, feeling sure the beast was about to land on me, straining every nerve for any sound which might give it away. Eventually, being still in one piece and unattacked, I plucked up the courage to fumble for the torch. There was nothing to be seen. Not even under my hammock!

Next morning the old hands told me it was probably a civet cat, a quite harmless denizen of the jungle, not very big as jungle cats go, but a lot bigger than your average moggy.

Every time we came out of the jungle we went on the ranges to fire off all the ammunition we had carried on ops. Some of the men were really good with their personal weapons. In fact, all were good, but some were brilliant. Keith, in particular, who was a good leading scout, always carried a pump gun (pump-action twelve-gauge shotgun). On a rough thirty-metre range one day, we were walking in turn across a rickety plank over a lot of mud and, somewhere in the middle of the plank, someone would give the order, we would turn to our left and fire at four 'figure' targets. In the case of the shotguns they were firing one round at each target.

Keith's turn came. At his first shot his hat slipped off the back of his head, and rolled slowly down his back. He hit all four targets dead centre then caught his hat with a hand behind his back before it could fall in the mud. If I had done that I would have waited for the applause, but Keith just walked on as if nothing had happened.

On operations our weapons were stripped and cleaned thoroughly every night. All the woodwork was taken off, cleaned and oiled with linseed oil. The weapons were never more than arm's length from us until they were handed back to the armoury after an operation. We never washed with soap, shaved or brushed our teeth on operations, as the smell of soap or toothpaste carries for miles in the jungle. Wading rivers and marching for hours in pouring rain kept us clean. It rained every evening, so if we felt dirty we stripped and stood in the downpour to get a shower. Our weapons, on the other hand, were spotlessly clean and gleaming all the time.

Before leaving the jungle we had to shave off our beards, put on a clean shirt and slacks and try to appear a bit 'soldier-like'. After one op I retained a big bushy moustache. When I got home Ann ran away from me, saying, 'Keep away, you look like Uncle Fred.' Poor Uncle Fred had been dead ten years, so I wasn't flattered enough to keep the shrubbery, and have never tried since.

Soon after she arrived in Malaya, Ann was very startled when she saw the local grass cutter at our window one day. She had never seen such an apparition. He was a Tamil, and very black. Like most of his compatriots he chewed betel nuts, so his mouth – teeth and all – were bright scarlet. He had a huge grass-cutting knife over his shoulder and had no shirt on. He just appeared there – while we were having breakfast – grinning from ear to ear.

The aborigines who came on ops with us as porters carried their own food, which they obviously liked. Their food seemed to consist of rice and a lot of small, dried fish which stank to high heaven. Being in the tail-end patrol on troop moves I often marched for days in the steady hum of *ikon bilis*, which is what they called it (small fish).

On my first op with 'D' Squadron my basha-mate shocked me by swapping a perfectly good tin of bully beef from our joint rations for a fistful of *ikon bilis*. I thought the heat had got to him – or the smell! He was an old hand, a damn good jungle cook. From that handful of stinking little fish he made one of the best meals I've ever had. After that meal, the smell of the porters' rations never bothered me in the same way. Towards the end of every march, when our one daily meal was nearly due, I found the smell positively mouth-watering.

Late in 1958 'D' Squadron went on what was to be its last operation in Malaya. None of us knew it at the time of course. There had been ideas of some squadron personnel taking their exams for the Army Certificate of Education First Class. To this end we were taking a lot of material into the jungle with us so we could study at night instead of reading books. We were on the trucks, having just drawn our weapons from the armoury, all ready to go, when someone called me off the truck and told me to report to the squadron office.

The squadron commander was there. He told me I would not be going on operations, I had to study for my First Class Education Certificate. I said I could do it in the jungle, but there was to be no argument. The squadron went and I stayed. Ann was surprised to see me back home. She had only seen me off a few hours before and didn't expect me back for about four months.

The exam was in three weeks' time and there were three of us from the regiment lined up for the exams, and four days a week we went to school again at an Army Education Centre in a camp near KL.

Finding myself with no possible reason to skive, I decided to report to the Medical Officer with a problem which had worried me on ops. I had noticed all physical exertion became more difficult towards the end of a long op. No one else seemed to deteriorate physically as I did, so I felt something was not right. I had never had to slow up on ops but it became progressively more difficult to keep up the pace.

I was given a thorough check: chest specialist, X-rays, the works. Nothing. The SAS MO, Captain Adlington, however, never gave up. He had me in again for yet another check-up. This time we went down to basic diet on ops. He had me list everything I carried in my rucksack. In ten minutes he solved the problem. I carried no sugar whatsoever. I didn't eat sweets or chocolate. I drank Oxo, never tea or coffee. I ate curry and rice, cheese and hard-tack biscuits. Nothing else. After that I took good care to have a brew of tea or coffee with sugar every day. When Ann found out she wouldn't let me have tea or coffee without sugar again.

Just before the exam, the squadron all came back to KL. There was something new coming up, but no one knew where or what. A squadron doesn't suddenly get brought out, cutting an operation short, for nothing. The exam came, and I passed the required subjects. The squadron was training hard in the meantime. The week after the exam we moved out of KL, down to Singapore and on to an aircraft for a destination unknown. There was much speculation about which war we were going to.

The Chinese Nationalists were in some aggro over the offshore islands. But that area was discounted as we could see no possibilities for us in that situation. A couple of places in Africa were gently simmering. Again nothing seemed to fit. After we left Singapore we knew by the direction of our flight we were not going further east.

We landed at Colombo in Ceylon, changed aircraft and took off again, flying north-west. A couple of hours in the air and the word came round we were heading for a place called Muscat and Oman. No one seemed to know much about the place. That seemed strange to us, with all the travel experience in the squadron. We landed at Mesiera, an RAF base on a small island, and changed aircraft again, this time to a Beverly, which could land and take-off on short, rough landing strips.

9 NORTHERN OMAN

So to the Sultanate of Muscat and Oman, an Arab country on the neck of the Persian Gulf, opposite Persia and Pakistan. The Beverly thundered its way over strange jagged mountains and bare open desert to land in a cloud of dust near Bait-el-Falage, an old whitewashed fort and a few mud huts.

In those days Oman was a long way behind the times in almost everything. It was nothing like the modern, bustling state it is today. There was, I believe, only about one mile of tarmac road in the country. All other roads were dirt tracks. I don't remember seeing any motor vehicles, apart from our military trucks and a few belonging to the Sultan's Armed Forces. All transport, as far as I could see, was by camel or donkey. There were no schools, hospitals, telephones, televisions or radios in the country, as they were forbidden by the Sultan.

The scene anywhere in Oman was just as it had been in biblical times, except for one very apparent detail: almost all males aged twelve or more carried a rifle of some kind. Many of them were old Martini-Henry rifles of .45- or .50-calibre, supposedly very good at long range. There were also quite a lot of Enfield .303 rifles about, the same weapon as used by the British Army until a year or two previously.

The men wore mostly white clothes, as per biblical times. The women were covered from head to toe in black, sometimes relieved by the odd touch of dark red. Only their hands and eyes were ever seen by anyone outside their own family. Most of them wore heavy silver jewellery, bracelets, earrings and nose rings. Most of the men wore a heavy cartridge belt, usually full of ammunition for their rifle, with a huge curved knife (*kunji*) in a heavy, ornately embossed silver sheath at the front, low, and in the centre of their stomach. The richer or more powerful the man, the larger and more decorative the *kunji*. I don't remember any local currency other than Austrian 'Maria Theresa' dollars, large silver coins. How they came to be in use in Oman I have no idea.

Our camp, which had been partly set up already by an Engineer detachment, but was still practically nothing, was about ten miles north of Muttrah, a sea port. Muscat, the capital, was a little further south down the coast, two or three miles beyond Muttrah. The buildings in both towns were not, I believe, of more than two storeys. They were of mud and rock construction, with flat roofs and small windows, many without glass but with wooden shutters. The streets, except the main through road, were very narrow alleys, being only two or three yards wide.

The old whitewashed fort, about a quarter mile from our camp, was nothing to do with the military scene. I've no idea of its use at the time, but the plain red flag of the Sultan flew above its tower. (The rebels of the day were said to use a white flag – very handy if you are not sure whether or not to surrender!)

Our first job was to erect a few tents in which to sleep and eat. We also put up a perimeter fence of rolled barbed wire. Then a squadron briefing. The squadron commander, Major John Watts, gave us a quick rundown on the country and why we were there. It appeared there was a British agreement to aid the Sultan any time he asked. He had asked.

It was the usual internal problem, with the rebels being armed and trained out of the country, then coming back to stir things up. Enemy strength was a bit vague, it was estimated at between two thousand and five thousand. Of which there were five hundred well-trained troops in uniform, the remainder being mountain tribesmen, well armed, with a savage reputation. A British infantry unit had had a go at them and came off second best, but that didn't mean a lot, knowing the infantry. The enemy held a big lump of high ground called Jebel Akhdar, a plateau with mostly sheer cliffs all around it, sitting at about six thousand or seven thousand feet, with parts rising to ten thousand feet or more.

It may seem strange to anyone outside the SAS, but it is a fact that after we were given the 'gen' on the enemy situation, there was not one of the forty-odd gun hands present who doubted our ability to crack the job. The squadron was by no means fully up to strength, and reports I have seen mentioning the 'sixty men' of 'D' Squadron forgot to mention the sixty included cooks, drivers, clerks, stores people and signallers. I was with 16 Troop and we operated with 17 Troop, which I well remember gave us a combined strength of twenty-one men (all ranks).

It is likely that 16 were the strongest troop in the squadron at that time with thirteen all ranks. 17 Troop, on the other hand, was a bit short-handed with a few away on other business. The thirteen other ranks of 16 Troop included three national service soldiers (conscripts). They had passed selection, para training and jungle training, and had been on operations in Malaya with the squadron. Compared to most of us, they were only short on experience (which was soon to be put right). As men who had proved themselves they were accepted by the rest of us without question. They pulled their weight, were no problem to us and they were a credit to the unit. In those days every SAS operational squadron had a few National Servicemen scattered among its troops.

The weapons in 16 Troop at the commencement of operations were three 7.62-mm Bren light machine-guns, nine 7.62-mm FN semi-automatic rifles and one .303 Enfield No. 4 bolt-action sniper rifle.

We each carried several grenades. I carried an FN rifle, six 36 and two 80 (WP) grenades, one hundred rounds of rifle ammunition in five magazines and two or three loaded spare magazines for the patrol bren-gun. We all carried roughly the same. However, being a supreme coward and having never run low on ammunition – but knowing many who had – I also carried two hundred rounds of spare ammunition in addition to the normal issue. I had adapted the lid of an old Bergen rucksack to use as a pouch, slung over one shoulder, for the extra ammunition.

Unusually for SAS troops at that time, we also carried our bayonets, as the flash hider for the FN rifle was incorporated in it. We had one radio in each troop. They were 128 sets, with no voice capability, morse only and, of course, we had to use secure codes.

Information on the enemy was not very exact: two thousand to five thousand, that leaves quite a big question mark. Five hundred 'well-trained, uniformed troops' also leaves a few questions begging. Like, how well trained and by whom?

'Someone in Saudi Arabia' was the only mutter I heard. So that left a lot of scope!

They seemed to have a reasonably impressive armoury, but again it was all rather vague: a 'lot' of light machine-guns, (most likely .303 Bren-guns); a 'lot' of .50 Browning heavy machine-guns; a 'few' medium mortars of unknown calibre or origin; 'plenty' of mines – both anti-tank and others. All were armed with rifles, either Enfield .303 or old Martini-Henri .45 or .50 calibre.

The only things which seemed an absolute certainty was that we were going to face an enormous amount of severe uphill, with a very target-rich environment somewhere along the way.

Once we were settled in and organised at Bait-el-Falage, we started training again. I had done no marching for a bit, due to the education caper, so I found it hard going. Mostly, though, we had to get switched on to shooting at distances of more than thirty yards. A lot of ammunition was used, but it was well worth it.

About a week after our arrival in the country we set off to get on with the job. 16 and 17 Troops went to the north of the Jebel, 18 and 19 Troops went to the south side. As we on the north side were rather out on a limb and the furthest from any other friendly forces (or even civilisation in the normal sense of the word), we were accompanied by the SAS Medical

Officer, Captain Adlington. Treatment for any injuries we were likely to receive would otherwise be a long time coming. Our troop medics were not as well trained in those days as they are now. Patrol medics didn't really exist. Captain Adlington climbed the mountains with us and was always on hand for any eventuality. For a couple of months of dodgy patrolling and being shot at on a regular basis, the Doc, and what he carried with him, was all we could expect for probably several days after any injury. After about one hundred miles of driving on very rough tracks, we arrived at the end of a wadi (valley), at a small village where the trucks could go no further. All around were great cliffs, rising thousands of feet. Our troop commander, Captain Rory Walker, chatted up the villagers and managed to get a guide who claimed he could lead us to the top of the mountain. That night we set out with the guide, marching in light order, to have a go at the mountain. We expected to meet opposition, and by the time we were nearing the top I was praying someone would start a war, so I could drop down for a rest. I was shattered, all that classroom time caught up on me.

Somewhere near the top we came to a sheer cliff of about two hundred feet and found steps had been carved in the rock. We heard later they had been carved there by the Persians, two thousand years ago, when the mountain had last fallen to an enemy force. Not long after the steps we found ourselves on the plateau and not a shot had been fired. We had caught the opposition off guard. The guide went down on his knees and gave thanks to Allah. I think he was the most surprised to have reached the top.

Marching had kept us warm, now we realised how cold it was at seven thousand feet. There was still a lot of time before daylight, three or four hours in fact, so we posted sentries and went to sleep. At least, we tried to sleep, but there was a strong, icy wind which didn't help; nor did the fact that we were in tropical shirt and slacks. Luckily, we had carried our Dennison smocks (camouflage paratroop smock) and these now saved us from some of the cold. Mine was a bit small for me, but I got legs and all inside it that night. Some of us had ice in our water bottles by morning. Daylight came at last, soon to be followed by a blazing hot sun, for which we were truly thankful.

Jebel means mountain, Jebel Akhdar means Green Mountain, but don't get the idea we found grass or trees on that plateau. There was very little greenery to be seen anywhere, except for odd bits of very low shrubbery, less than knee height. There were more camel thorn trees than are usual in the desert, but these were not plentiful, or even big.

The plateau was not flat by any means. It had its own hills and valleys, cliffs and open spaces. The hills on the north side were often several hundred feet high, easily defended but easily shot at from good positions on other hills all around. The valleys varied from shallow depressions with gentle slopes, to huge cracks in the plateau with sheer rock cliffs going down several thousand feet, almost to sea level. These cracks were caused by water over the centuries and were sometimes one hundred feet wide, sometimes half a mile.

The plateau was about twenty-five miles long, its width varying from a mile or two to a maximum of about ten miles. Like most mountains in that part of the world it has a predominantly rocky landscape. Rocks of all sizes everywhere, from the size of gravel to ten-storey-house size. There was sandy earth on most of the flat areas but it was well cluttered with rocks, some loose and some embedded. The rocks were all shades, even multi-coloured. A man could lie down almost anywhere and not be seen from twenty or thirty yards. On the other hand, a man walking could often be seen from several miles.

As with most areas of the Middle East, an arid, dry landscape is ensured by the daily deployment of vast herds of goats which nibble away anything showing the slightest hint of green. Whether or not the herds of goats and their child shepherds we saw near the top of Jebel Akhdar returned each night to their villages, some eight thousand feet below, I have no idea. But if they did it's a mystery where they all got the energy.

Before going up on to the Jebel we were told it was 'green' and there was fruit up there. It was a standing joke among the trogs that the only fruit we saw on the Jebel was a certain officer (not SAS) who came up much later in the campaign.

As soon as it was light we went off in patrols to check the immediate area. No contact was made with the opposition. The area seemed dead. It was then decided we would stick to what we had taken so easily and extend our foothold as and when we could. To this end, 17 Troop headed back down to collect their packs, we could obviously not exist without our full kit and rations, especially at that altitude. There was the small problem, of course, that it would leave 16 Troop on their own until the next morning. Ten or eleven men had their limitations against an attack by possibly one hundred or more.

Anyway, nothing ventured, who dares and all that, we looked to our weapons, did what we could to improve our position and made the most of the sunshine. By midday the sunshine was getting a bit much and we were looking for shade.

We were wearing khaki drill trousers and, on looking around, I decided they were definitely the wrong colour for the terrain. Something would have to be done to tone them down a bit. Our smocks were ideal, so at least we were halfway to being bad aiming marks. An old soldier once told me, 'Half the battle is not to be seen, the other half is not to be easily seen through a gun sight – you can't always achieve both, but if you don't get one right you're a cert for the boneyard.'

Towards evening a couple of patrols went out but found nothing. Before dusk we built a huge fire to help us through the night. We put the fire in a small hollow so that after dark the embers would not be seen from anywhere around. Darkness fell and we closed in around the hollow, nothing had been seen all day, we felt reasonably secure. It was important that we get as much sleep as possible while we had the chance, so we lay in a close circle around the hot embers, with one man on sentry, and all within arm's length of the next man.

My turn on sentry was around midnight. I moved to the machine-gun, which was the sentry post, and began to study the shadows and darker shadows which made up the local scenery. After about twenty minutes some shadows started to move. I looked away immediately, checked other possible danger points, then looked past the shadows which had changed. They had moved and were still moving. Again I checked all other possibilities, again nothing.

Only one area was wrong. Bringing the machine-gun to bear on the approaching danger, I kicked the nearest sleeping man who immediately rolled over into a firing position. Seeing I had started the 'stand to', I watched the shadows, which were now an approaching mass about fifty or sixty yards away. A brand new sliver of moon glinted off a rifle barrel in the approaching unknown, so I slipped the catch to automatic, practised a low sweep across the front of the mass and took up first pressure on the trigger. By this time the whole troop was lined up on the target all ready to roll it out. All except two that is, who were watching the opposite direction, in case other shadows started to move.

Being the sentry and having the machine-gun lined up, it was my ball to open, but something was wrong. It was all too easy. No attack would come in a heap of a target like this. Captain Walker thought the same obviously, when the oncoming crowd was at about thirty yards' range he whispered, 'Hold your fire,' and instructed the Arab guide to challenge them.

At the first shout the mass stopped dead. One small sign or sound of weapons being prepared and we would have wiped them out. The Arabic went back and fourth a few seconds, then a quick interpretation. Captain Walker said, 'It's okay, they are this bloke's people.'

We had been instructed by the Sultan of Oman to kill everything on the Jebel. There was nothing up there except enemies. Kill every man, woman, child, camel, donkey and chicken. The people obviously knew about these instructions and thought we would carry them out. The 'mass' turned out to be about a dozen donkeys being led by four men. They were very surprised to find us there and, as they were really there illegally, being forbidden to travel over the mountain, were very grateful their lives had been spared.

With the help of our guide we soon made friends, and found that these tribesmen regularly used that route over the mountains as it saved two or three days' hard travel between their village and the nearest town. Captain Walker quickly saw the possibilities of this chance encounter and enlisted the help of the men and their donkeys to bring up supplies until something better could be organised. Without those people and their hardy mountain donkeys we would have been very lucky to have retained our flimsy toehold on the plateau in the early stages.

No one in Middle East Command had expected us to get to the top of the plateau so quickly, so there was no supply system prepared in advance. As usual, things were to come down to the sheer luck and ability of the man on the ground. A signal had been sent to Squadron HQ telling them we were on the plateau. Major Watts didn't question the signal, but when he passed on the information to HQ Persian Gulf they asked him to check again. I have heard John Watts then replied to the effect that if his troops said they were somewhere – then that's where they were.

In the morning 17 Troop reached the top with all their kit, a Gloster Meteor reconnaissance jet flew over and around our position, having a good look, presumably to check we were really where we said we were.

The first two or three days at the top we managed to find a small amount of water in holes in the rock. When that ran out the donkeys came into their own, bringing water and rations up the mountain.

After 17 Troop reached the top with their kit, 16 Troop did the trip down that day, then spent the following night flogging slowly up again. It was a hell of a climb and took all night. On the climb up with our kit, 16 Troop had the dubious job of escorting a small donkey train; about six friendly tribesmen were each in charge of a donkey which was heavily laden with rations and jerrycans of water. Half a dozen of us were each allotted a tribesman and donkey, with orders to make sure they reached the top with the supplies intact.

We started the climb in daylight about an hour before sunset. Just before dusk I found myself lagging further and further behind the column. My donkey handler seemed to be having a lot of trouble with the donkey and its load.

Twice the load had slipped around under the animal and we had to reload it. Several times the donkey stopped, stubbornly refusing to move.

By then I had realised my Arab companion was the village idiot. We couldn't speak to each other because of the language problem. Eventually the donkey turned around and set off downhill. Its handler stood and laughed insanely, so I dropped my pack and managed to catch the beast, bringing it back to where the handler stood, still laughing.

We had been warned not to antagonise the locals, we needed all the friends we could get. The thoughts of being diplomatic flashed straight out of my mind, even quicker than my knife came out. The point dented the skin of his throat just under his chin, I pointed at the donkey, then at the track upwards. I also probably looked a bit upset.

The village idiot laughed until he drooled. By this time he was up on his toes, and I was expecting to see blood. Reason mostly prevailed. My bluff had been called. The knife returned to its sheath, the laughter grew louder.

Reason didn't quite prevail. Unreasonably, I vented my frustration on the poor dumb animal, aiming a punch at its head. As I threw the punch the donkey happened to shake its head, so where it landed I don't know, but the animal went down like it had been shot.

The laughter stopped. The donkey struggled to its feet, looking in better shape than I had ever seen it. The handler stared at me open-mouthed for a few seconds, then without so much as a smirk he went to work on the load, putting it straight, tightening the girth. Within a minute we were on our way up the hill, the donkey more than willing and the handler pushing it, with occasional fearful, backward glances.

It was the luckiest punch I've ever thrown; hole in one; all problems solved. We soon caught up with the troop. That climb with full kit was a memorable, nasty, long, flogging struggle against gravity but at least we now had the bare necessities of life with which to face the odds.

Our sleeping bags were the big old-fashioned heavy duty, semi-waterproof, airborne things which weighed a ton, well, about twenty pounds I would think. They were carried strapped to the outside of our packs, usually underneath, and their bulk was almost as much problem as their weight, but we were damned glad of them during the cold nights of the next few weeks. Once both troops were really on the plateau with all our kit we started to push south and east. There was no point in trying to take on heavy odds on their ground. What we wanted was ground of our own and hope we could pull them on to it, so there was no mad rush to make contact.

We heard of our first casualty with some anger. 18 and 19 Troops on the south side had hit some opposition who had managed to kill Duke Swindells. Duke was well known and well liked. There was now a feeling that a mad rush to hit them might be a good idea. Rory Walker kept his cool – things progressed as planned.

The 16 Troop sergeant, Herbie, took a six-man patrol forward about two thousand yards to set up an observation post. They were there all day and just before dusk were attacked by thirty or forty enemy. The opposition came in a fast fire-and-movement attack but got their just deserts from Herbie's patrol, losing about ten men. Our patrol withdrew according to plan, but there was no follow-up. In the morning we all went forward, expecting to have to fight to retake the position, but found the opposition had had enough, not even taking the position the night before.

That small skirmish had a good knock-on effect for me. The extra spare ammunition I carried was about halved by replacing that used by the patrol. My belt-weight was lighter by about one hundred rounds of ammunition.

Captain Walker decided to move the whole set-up forward to a hill we named Cassino. It dominated a considerable amount of real estate and was readily adaptable for defence by our small force. We had just about taken up our fire positions on Cassino and were getting sorted out with our building programme for *sangars* (rock walls instead of trenches where it was impossible to dig) when I heard a sound which has always made me nervous, since April 1951: the hollow clang of a mortar being fired. Everyone heard it and all movement stopped. Someone said, 'It's only a "woozer".' Which was our nickname for the tribesmen's ancient Martini-Henry rifles.

I said, 'Shut up and listen.' Then we heard the shell, high above, swishing down towards us.

It exploded with a crash about two hundred yards south. The mortar coughed again. We were flat on the deck. This one exploded about two hundred yards north. We were bracketed. According to all my training and experience we were about to get a pasting. We waited and waited. Nothing happened. That was it. What sort of enemy fires a good ranging bracket – then stops. I thought I knew. An enemy who is waiting for the right moment, which meant he was waiting for someone else to get into an attack position. I felt quite certain the enemy was about to launch a full-scale attack. Happily, it was one of the many times in my life when I was wrong.

The same thing happened again the next day. Between five and six o'clock in the evening, two or three ranging bombs then nothing. We called it the 'Six-Five Special' – after a well-known UK television programme of the time.

They fired their two or three bombs at us almost every evening. After the first time, we went out to find where the mortar had been fired from and found the position. Also, we found – from bits of rubbish left at the position – it was an 81-mm mortar. This meant it was a better weapon than any we had available. The troops on the south side of the Jebel had our 3-inch mortar group with them, but the 3-inch mortar hasn't the range of an 81 mm. On the north side we had nothing except our personal weapons (FN rifles and Bren light machine-guns). We did have one .303 bolt-action sniper rifle in each troop, but they were, I thought, more a liability than an asset.

We asked for something with which to hit back at defensive positions and were sent some '94' grenades. These can be fired off the muzzle of a rifle by using ballastite cartridges, but, although quite powerful for their size, have a maximum range of about one hundred yards. There was also a big snag: the ones we got came with .303 cartridges, so were useless with our 7.62-mm FNs. At a push they could be used with the sniper rifles, if we were prepared to sacrifice any accuracy in future sniper work.

Quite often the opposition fired ineffectively at Cassino from long range. They were certainly not living up to their reputation of savage fighters, nor did they appear to be well trained.

The nearest the enemy came to causing us a problem was when they crept up onto what we called 'Herbie's Hump'. It was the position where Herbie's six-man patrol had our first contact. The Hump was about two hundred yards or less east of our nearest positions on Cassino, and although it was lower than our position we could not see over it. The side facing us was a sheer cliff of about thirty feet, the other side was a gentle slope with large rocks and bushes on it. The cliff was about forty yards long with sharply sloping ends. My *sangar* was at the east end of Cassino, facing Herbie's Hump and nearest to it. There were two of us in the *sangar*.

The second afternoon we spent on Cassino about a dozen or so enemy opened fire on us from the Hump with .303 rifles and two light machine-guns. A considerable amount of spiteful ammunition splattered among us. We were not on stand-to, just resting, so it caught us a bit off balance; no one had twigged the enemy's approach. I grabbed my rifle and felt for grenades as I tried to get a look at the attackers. There was a lot of fire hitting my immediate area.

Rock splinters and gravel showered down on us as the rounds struck all around like small explosions. Bits of rock stung my face and bare right arm as several rounds hit the rocky slope to my right where the *sangar* wall was lower. It felt like a good time to be wearing goggles – or even full armour.

Uppermost in my mind was the fact that the last time I had taken on this type of weapon at this range, I had come off a very definite second best. Nevertheless all experience is good for something (usually) and my experience told me this machine-gunner was a totally different kettle of fish from the cool, calculating bastard who had nailed me in Korea. His bursts of fire were too long to be accurate, and he sounded more like a man in a hurry to get rid of ammunition.

Another difference was the weapon I now held ready for action, confidence seemed to flow from the familiar feel of the lethal wood and steel of the FN. On top of that there were no exploding mortars or grenades. No scrambling, panting, wide-eyed tommy-gunners (I hoped!). I wasn't bleary-eyed, tired out, battered, or green.

He was mine for the taking, even before I'd spotted him. Easing up to peer through the dust and smoke I could see the black magazine showing plainly against the background of a light-coloured hat or *shemag* as he fired from between some rocks on the rim. The sights of the FN lined up on him, everything was right, I squeezed off the shot and missed. The puff of white smoke off the rock a foot or so to the left surprised me. It had looked so bloody easy. He didn't like it though, my second shot was too late to catch him.

The FN sights raked along the rim to the left, a slight movement, another shot, probably another miss, then the enemy fire ceased. I raised up to look for a target, there were none. I felt cheated, staring at the now empty enemy position with dust and smoke still hanging over it and knowing they were gone. Knowing I had taken my big chance for revenge and blown it. No excuses, I should have made that shot count. My great chance to get a bit even with machine-gunners who picked on me. Shit!

We had no casualties. Enemy fire had been heavy but wild, only lasting about twenty or thirty seconds before they bugged out.

At that stage of the operation we were not sure of the enemy's tactics or capabilities. The attack could have been an attempt to pull us off our defensive position in a follow-up of the attackers – only to fall into a classic trap sprung by a larger hidden force. Their knowledge of local terrain would indicate this type of tactic as a possibility. Later experience makes me feel there was no trap. They were just trying us, testing the water. We didn't bite anyway.

They did a repeat performance from the Hump an evening or two later, but it was over so quickly I didn't even get a shot at them that time. I wanted a clear target again but couldn't find one. Again there were no casualties on Cassino. After that the enemy confined themselves to longer range, even less effective fire and, of course, the 'Six-Five Special'.

A few hours after the first attack from Herbie's Hump, one of our people who had been away on a patrol somewhere to the west, arrived back on Cassino. Covered in dust and sweat, he thankfully unslung his heavy belt, laid down his rifle and said, 'It's all go, innit!'

A voice from the next *sangar* said, 'Don't bloody bet on it, mate, a couple hours ago it was all coming back!'

Had the enemy known there were so few of us I think our troubles may have been considerably greater, but we patrolled very actively day and night. The daylight patrols were either within reach of assistance from Cassino or were back along the Jebel to our original position and the supply route, ground which we called 'ours'. At night we patrolled anywhere, mostly forward towards the enemy positions. Even if the enemy didn't see or hear us at night their scouts and patrols must have found our tracks all over the place.

Although we didn't know it at the time, Radio Cairo was conducting a propaganda war against us, claiming the heroes on the other side were wiping us out in hundreds – if not thousands. So it is quite likely the enemy thought they were up against a huge military force. This would have been much easier for them to believe than the plain truth. Otherwise, even a couple of them with rifles could have closed down our supply line quite easily as we had no one to defend it. Cassino was virtually isolated, well within their reach and capabilities. We had no ground-to-air radio facility to handle close air support and there was no chance of a relief force getting through to us within several days (even had such a force existed!).

It would be quite unfair and untrue to say the tribesmen of Jebel Akhdar were useless fighters. They were many in numbers and there were undoubtedly brave men among them. Before we arrived on the scene they had a ferocious reputation. We knew that on one occasion they had charged down the lower slope of the Jebel to attack a position which contained artillery. The charge had been so fast that one of the guns firing at them had depressed its aim too low and destroyed its own emplacement. It had taken air strikes and armour to chase the rebels from towns on the plain, near the Jebel, and push them back into the mountains. Since then everyone who had tried to pursue them had retired with a bloody nose. There was quite an impressive casualty list.

They had shown guts enough at times but they were, after all, civilians. Good old boys who would carry a gun and have a go at anyone who had the temerity to get on 'their' Jebel. They could not, however, be expected to cope with the SAS, which had a high proportion of hardened, well-trained, experienced professionals who didn't give a damn for anything except doing the job. To add to the rebels' problems they were also led and advised by civilians. Badly led and badly advised.

We also had our problems overshadowing every move we made. The whole scene would have been closed down much quicker except for our 'softly, softly' approach to the operation, necessitated by strict orders to avoid casualties as a top priority. Yes, just think about it. Send men to do a job which, on the face of it, looks almost suicidal – and, at the same time, order them not to risk casualties. Sounds like the big political umbrella. Anyone who gets his men or himself killed must have disobeyed orders – therefore don't blame us!

The pressure was heaviest on our field commanders, especially John Watts as squadron commander and ultimately responsible in the field. Rory Walker also had the problem of often holding us back when experience and better judgement (his and ours) said, 'Do it!'

It was almost like trying to play a game of chess – or draughts – without losing a piece. There were many times when members of 16 or 17 Troop wanted to pull some kind of trick on the opposition. Seldom was anything allowed, although I believe an attempt was made to capture the enemy mortar by a small fast-moving night patrol. The idea being to catch them red-handed, so to speak, when they fired their evening shots. It didn't come off though. The enemy had enough sense to fire from many different places.

After a lot of studying of Herbie's Hump, I had the bright idea of using a small ledge on the cliff facing Cassino as a base from which to throw grenades among future attackers. The snag was, whoever did it would have to lie concealed there in the heat of the sun all day as there was no way of reaching it undetected in daylight.

Mentioning the idea to George started an argument between us as to who should try it. We had to fix it all with Rory, so decided he could nominate which of us went. The other would cover with a rifle from my *sangar*. Rory listened to the plan, thought about it for all of three seconds, then told us to forget it. 'The sun must have got to you already!'

To help with the supply problem some donkeys were brought in from Somalia. They were small and not much use compared to the big local donkeys which were almost as big as mules. They were only allowed to be loaded with fifty pounds, whereas we trogs – with only two-wheel drive – were expected to carry twice that and more. Some of us went down to collect a batch of donkeys and stores. One of the men got kicked while trying to feed a donkey. We started to go off them at that point. By the time we reached our base at the top we had exhausted ourselves, carrying our heavy loads and dragging the reluctant mokes.

We had strict orders that no donkey would be shot unless it went lame. Just before we reached our base there was the familiar double tap of two fast shots from an FN, and a donkey fell down dead. The man whom it had kicked stood there with a smoking rifle saying, 'It limped, the bastard limped, I saw it limp!'

Within half an hour the donkey had been neatly dealt with and we were turning some juicy steaks on improvised spits over several wood fires.

So we had our first fresh meat for several weeks. It was a great treat after all that time on tinned rations. In fact, our rations at that time were some of the worst I have ever had to exist on (barring PoW rations in Korea). All our rations were what the army calls Twenty-four-hour packs – they were not designed to be the only source of food for a period of several weeks.

To make matters worse we were getting only one type of pack sent to us, so we had no variation to our diet. The same food day after day for weeks can become very boring. I would have three or four biscuits with a small tin of cheese and jam from a tube in the morning, then put the whole day's ration into a sort of 'biscuit stew' in the evening. We had no rice or potatoes and no bread, so the few biscuits in the ration pack formed the bulk of our daily meals.

Soon afterwards, we got our first resupply by air. It was a fiasco. All the more so after being familiar with the near-perfect air-drops by the great combination we had in Malaya – RAF Valettas and Bristol Freighters with the men of 55 Company RASC handling the drops.

This was totally different. Our DZ was not a wee hole in the trees which needed a lot of finding, but a damn great patch of open ground big enough for ten football pitches. The aircraft was a Hastings and it dropped its load from about two thousand feet, when it should have been down to five hundred feet. Those parachutes which worked, drifted off into the distance and, once over the edge of the plateau, were lost to us forever. The packs which broke loose came down like bombs all over the place. Nothing in them was usable.

A few days later we had another drop. In the meantime one of our mortars was being sent to us from the south side, so we had ordered mortar ammunition. The drop came in lower this time, about a thousand feet. The packs, however, broke away in great order. Having seen the first effort we were all well clear of the DZ. A good thing too!

Whoever packed that lot had really worked on it. The High Explosive (HE) mortar shells were in the same packs as the phosphorous shells. When those packs whistled in the phosphorous shells were smashed, so the phosphorous set fire to everything all around. The HE shells got hot and began exploding when they felt like it.

We salvaged what we could. A few brave souls rushed in and out of the exploding DZ trying to get the mail, the decent rations (the rum ration?) and anything else they could. Like most of the others, I got behind a good size rock, lit a cigarette and waited for the shrapnel to stop screeching past before venturing out into that lot.

I don't know what signal Rory Walker sent about the air-drops, but something sparked someone, somewhere. After that disaster, things improved rapidly.

Having located the enemy's main defence line, Captain Walker decided to try out our other air support, a squadron of ground-attack fighters based at Sharjah, in what is now the United Arab Emirates. We had no ground-to-air radio links at that stage, so things were a bit tricky once the aircraft were overhead. Enter now a clown, yes, even in the SAS (did I say 'even'? At times it was full of them!). The clown had already proved himself by eating a whole troop's bread ration 'by mistake'. That wouldn't have been so bad but it was the first bread we had seen (or didn't see!) for several weeks.

We had received an air-drop earlier and the marker panels for this had supposedly been brought in by the clown. We had laid out the marker panels for the air-strike just in front of our positions, on the north-east side of Cassino. The panels were in the shape of an arrow, pointing in the direction of the enemy position we wanted to be hit. The jet pilots would have been told the bearing from our arrow to the target and the approximate distance.

We were having a brew of tea when the jets came on the scene. They were Venoms – a twin fuselage jet fighter, the Naval version of the more common Vampire. We watched them circling, getting the lie of the land. Then I saw something was wrong. I was likely the only one there who had been attacked by jet fighters before, so I recognised the pattern of flight. They were stacking up to hit Cassino.

I yelled a warning to the others and grabbed a towel, or something white, stood up and waved like mad. The men around me followed suit quickly. I watched the rocket pods on the first aircraft. The first puff of smoke would see me in my *sangar*.

It never came. The two Venoms screamed over our heads and away. We heard afterwards the leading pilot couldn't believe his eyes when he saw – with his thumb on the button – his target waving at him! Realising there had been a bog-up, the Venoms returned to their base without firing a shot.

We looked to our set-up for the reasons and found them. The sun was dropping in the west, our arrow on the north-east side of the position was now in shadow and not easily seen from the air. It was an easily made mistake which could have cost us. But there was worse: the clown had not collected the air-drop panels. The wind had blown them into a rough arrow pointing at Cassino, and the bearing from those panels to us was almost the same as from us to the target.

We still had a lot to learn, but would we have time at this rate? Rory

Walker made sure we all knew the whys and wherefores of that lesson. I know it's a lesson I will never forget.

To help with our fire-power problems some .30-calibre Browning machine-guns were borrowed from the Lifeguards. They had plenty to spare, as most of their armoured cars had been blown up on the roads and tracks around the foot of the Jebel (by mines originally supplied to Saudi Arabia by the Americans!). Also a couple of 3.5-inch rocket launchers were brought up to our position. I think we had two or three .30-calibre Brownings and a couple of lads from the Lifeguards came with them to help us out. With these extra weapons an attack on the enemy's stronghold now became a better proposition. A company of infantry from the locally raised Northern Frontier Regiment arrived to defend Cassino, giving us a base to work from and releasing both troops for offensive operations.

The enemy positions were about four thousand yards from Cassino, so a 'step-up' operation was needed to give us plenty of time to cope with them during darkness. Halfway between us and the enemy was a ridge we called Middle Wallop. They had fired at us from there with rifles and machine-guns. Their 81-mm mortar had been positioned behind it on several occasions, to fire its few evening shells. To the left of Middle Wallop and a bit further from us were twin peaks we called Sabrina (in honour of a certain well-known TV personality). The enemy often patrolled Sabrina and I believe they had a permanent observation post (OP) somewhere there.

The enemy's main defence line was a natural fortress: a cliff about fifty feet high, maybe more, which barred the way to the main plateau. There was a gap in the middle of the cliff, about twenty-five yards wide, partly blocked by huge boulders. It was obviously the front door to the fortifications, heavily guarded by at least fifty enemy who were in the honeycomb of caves all around it. The proper name of the fortified cliff was Akhbat el Zhufar. We called it the Akbat.

We spent a lot of time scanning the whole area through binoculars, watching enemy movements and trying to find where there were obstacles barring our way. It was very difficult to assess the ground at any distance as it was so broken and disrupted with jumbles of loose rock in piles many feet high and deep fissures where they were least expected. Some of the wadis were over a mile deep, with sheer cliffs from top to bottom.

The view from Cassino was vast. On a clear day, which was most days, we could see right across the coastal plain to the north-east, right out on to the blue water of the Persian Gulf. Through binoculars we sometimes saw ships out there and envied their crews with their good food and plentiful liquid refreshment. The coast was about fifty miles away. To the south and west the sands of the empty quarter stretched away into infinity. It looked very empty at that distance.

The main part of the Jebel was roughly east of us. Almost every day heavy bombers, RAF Shackletons similar to the famous Lancaster of the Second World War, would appear at a great height and drop huge bombs on one particular area, miles away along the top of the Jebel. Using our binoculars we could see the bomb leave the aircraft and watch it fall to explode with a great pall of dust and white smoke on what appeared to be a bare area of mountain top.

There was much speculation about what the Air Force were trying to achieve. Some thought it was just a bombing range. Others thought the target was Saiq, the rebel headquarters. I remember a discussion when one voice said, 'It can't be Saiq they're bombing or we would be able to see the houses where the bombs are landing.'

Another voice answered, 'Not if they've been dropping those bloody things on it for the last fortnight.'

In fact the bombers (we were told later) were trying to destroy an area where the rebels were supposed to be storing a lot of military hardware in huge caves. Also the rebel leaders were supposed to have their headquarters there.

In case of casualties we had to find a place where a helicopter could land and – most importantly – take off again. The one ancient Sycamore available had a casualty-lifting ceiling of seven thousand feet, and I remember being on a patrol to check on LZ situations about two thousand feet below where we were operating. So I presume we were at about nine thousand feet on Cassino.

On one patrol a swarm of locusts came upon us and we had quite a time covered from head to toe in the damn things. They seemed to like the look of our smocks and KD. My KD slacks were now covered in patches of black boot polish to cut down the glare, but it didn't stop the locusts.

All of us had badly chapped lips from the wind, the cold nights and the burning hot sun by day. Our boots, which had rubber soles, were worn down so that the screws which had held the soles on were like football boot studs with rubber washers under them. The toe caps were mostly worn away and some of us had our toes showing through. After six weeks, a resupply of boots was most welcome. To say the going was rough was quite an understatement.

We were all impatient to have a real go at the opposition. Our respect for their much-vaunted abilities and superior numbers had dwindled to zero. It was high time they were given a short sharp lesson.

10 'AKBAT' AND TANUF SLAB

Captain Rory Walker worked on all the information we could gather from patrolling and observation. He made a plan, put it before the usual SAS Chinese Parliament – so we could all get our cribs out of the way at the start – then, with a few changes to allow more or less time for this and that, we went ahead with it. As with all good plans, there was scope for changes as it went. We all knew what to do in the case of this or that going wrong. In the event so much went wrong and so much went right our plan worked anyway.

It was decided that 17 Troop would make a diversionary attack in the area of Sabrina. 16 Troop would scale the cliff of the Akbat and hit the main strong-point from the rear. The overall plan being to badly shake or destroy enemy morale. There was no question of it being more than a raid, as we had not the men to hold such a large feature and fortification, even if we could take it.

As we had no radios, apart from the morse set to Squadron HQ, everything depended on timing for support and diversionary fire.

At the time we were hoping to be scaling the cliff, the machine-guns were to engage the enemy positions to cover any noise we might make. A rocket launcher would also be used to hit the main strong-point just as we reached the cliff.

As we had no real maps of the area (Captain Walker had the one map, which was a white sheet covering nearly all of Northern Oman) route planning on the final approach was impossible. There was so much dead ground (real estate out of sight) between us and the Akbat that we had to make do with more hope than certainty.

On the first night both troops and the machine-guns moved forward to take Middle Wallop. There was no opposition but we did almost get entangled with a big enemy patrol (which had fired at Cassino, now held by the Company of Omani Northern Frontier Regiment). We heard their running feet a few yards away in the darkness as they headed back for their defence line.

Before daylight, 16 Troop had moved forward to a position directly in front of the enemy's 'front door' and about six hundred yards from it. We lay in the open all day, watching the enemy and hoping their patrols didn't stumble upon us. As soon as it was dark we started the final approach, leaving our signaller and the two-man rocket-launcher crew to do their work from that position.

About four hundred yards was covered quickly and fairly quietly. Then we came to a deep ravine. The ravine had steep sides and was about eighty feet deep. It took some time to find a way down. So long, in fact, we knew our timing was going to be out. As we reached the bottom and contemplated a stiff climb up the other side we heard 17 Troop's diversionary attack going in to the north.

At first I didn't recognise the sound of firing. The sound was strange and startling. Perhaps because we were in a deep ravine the crash of rifle and machine-gun fire, echoing from a thousand rock faces, passing high above us in the night, made a weird swishing sound. It started slowly, swish, swish, swish, then built up rapidly into a long, shuddering ssshhhhh, like a strong, fast approaching wind. Eerie.

Then, just as we were starting to climb the far side, rockets started exploding just up the ravine. We realised they were our own rockets, falling short by two hundred yards. Some rethinking had to be done. The timings had to be altered. Captain Walker gave me a message to take back to the signaller and also a verbal message for the rocket-launcher crew.

The night was inky black, except for a few stars, but I knew I could find the three we had left behind. The only problem was, they weren't expecting anyone. I hoped they were all together and one hadn't wandered off a few yards so I'd bump into him before schedule. All went well, the timings were rearranged through SHQ to 17 Troop signaller with the machine-guns.

I galloped back to the ravine and we all started off once more. Soon after leaving the ravine we came to the Akbat cliff. We had spotted a place which might be scaled and moved along to it.

Captain Walker, who always insisted on being leading scout on troop moves, tied the rope around his waist and started to climb. The first thing we noticed was that the rock crumbled under his feet and hands, making for a tricky climb. About fifteen feet up he was stuck, couldn't go up, and had to be helped down.

After a quick whispered discussion it was decided to move north, to try nearer the main strong-point. I volunteered to recoil the rope while the rest quickly moved out. The rope was snagged in the rocks and took a few seconds to organise. With the rope safely hung around me I took off after the troop. Ahead of me I heard a few shots, nothing came my way so I kept going along the bottom of the cliff.

That reorganising of the climbing rope, although taking valuable seconds to complete properly and causing me to be suddenly alone in the dark on the very spot where two short hours before, through binoculars, I had seen a twelve-man patrol of the enemy moving slowly along at full alert, was one of the best moves I've ever made. Not only was it coiled so as to be ready for its intended purpose but its tidiness ensured my spare magazines and grenades were still at maximum availability. A fact to be thankful for soon after. At the base of the cliff there was a well-worn track which, although undulating in places, was fairly free of loose rocks, making little problem for quiet, fast movement by starlight.

Sometimes weaving between huge boulders, sometimes in the open, I rushed on for what seemed to be quite a distance, all the time expecting to catch up to the troop who, I thought, would be waiting for the rope. It did cross my mind there was a possibility of catching up with the wrong people, or perhaps meeting them. Trust in one's comrades, their expertise in this type of work and our codes of practice become paramount on such occasions.

By the time I caught up, I was beginning to think I'd missed them somewhere. I asked where Rory was, thinking he would need the rope. I was directed up between some rocks and ran quickly up through the gap. Coming out into an open space I heard the unmistakable sound of a grenade being thrown. I immediately hit the deck, not knowing who or what was where.

The grenade bounced and clattered on rock then exploded. I heard Rory say, 'Give them a phosphorous!' A couple of seconds later I heard the sound of the phos being thrown. When it exploded I saw it light up a cave entrance, then all hell broke loose.

I saw Scouse, one of our Bren gunners (not he of the leeches), stand up in front of me, so I joined him. We stood about six feet apart, firing back at every gun flash. Most of the fire was coming from above us. Although feeling sure I would be hit – there seemed no way they could all miss – there was no time for the luxury of fear, anger or other straightforward sensations, we were far too busy. Realising whoever else of our people was in there with us was about twenty yards away to our right, I knew that everywhere else was open season.

Scouse raked the thickest areas of muzzle flashes with automatic fire; I fired single shots at the nearer enemy muzzle flashes, taking the ones at about twenty to thirty yards first when possible. At least two were only about ten yards from us but they only fired one or two shots each before the return fire closed them down.

At no time during the raid did I see a clear target. All one sees during a night action, once firing commences, is muzzle flashes and grenade bursts. The trick is to keep one eye closed for as long as possible – if it's a short sharp job – to prevent temporary blindness caused by the gun flashes. On this particular occasion, being right in among the opposition, it was another luxury I couldn't afford. Blurred shadows, a couple of half-seen, split-second silhouettes and a glinting cartridge belt is all I remember.

A shot hit the Bren magazine and Scouse did the fastest magazine change I've ever seen. I went down on one knee, to make a smaller target, and noticed the return fire had slackened.

A shot came from my left. I don't know where the bullet went, all I noticed was the muzzle flash about four or five yards away. This could have been the one which had hit the bren magazine with an earlier shot. Then I saw it was a cave or hollow at the foot of the cliff, with a four-foot rock wall at the front of it. I put a quick double tap into it, followed by a 36 grenade over the wall and two or three more shots spaced across the darkness before the grenade blew. After that there was a sudden shortage of targets. I fired a few shots up at the cliffs where most of the enemy fire had been coming from, then heard Herbie calling us off. It was he who had been with Rory, over to our right.

Scouse and I were only too glad to leave by then, as everyone else seemed to have gone – opposition and all.

Well, most of the opposition. In the comparative silence which followed the gunfire there arose those never forgotten sounds of such places: sounds of choking, groaning, retching and babbling; voices of men crying out in agony and fear; someone thrashing about on the ground a few yards away, making a gasping, bubbling sound.

Over to the right, where Rory's voice had called for the phosphorous grenade, another vaguely familiar voice now sounded. It was pleading and incoherent. I stopped in my tracks, then started back in the direction of that voice. After a few steps I stopped, listening, then called the man's name. No answer, so I started to go further. Herbie rushed up behind me and grabbed my arm, 'Get to hell out of it, Lofty!'

Still not sure, I said, 'Are all ours out?'

'Yes,' he said and gave me a shove in the right direction.

Good old Herbie, always on the ball. He always knew where everyone was, even in the midst of chaos. Four of us had been in the middle of that lot, and I didn't even know what was going on – didn't have an invite. Not a scratch between us, it was quite remarkable. For comparison it was like standing in a street of tall houses and being shot at from all the windows at once.

As we pulled back with the rest of the troop I saw a big cave at a lower level, and suggested a grenade would make it safer to turn our backs on.

Rory agreed and said, 'Make it a phosphorous, though.' So I did. It wasn't a cave but a great stack of brushwood. The grenade turned it into a huge bonfire which lit up the whole area.

It was only then I realised where we had been: into the front door of the great Akbat strongpoint. I should have known – it was the only break in the cliff.

Now, thanks to that last grenade, we walked away from our handiwork lit up like broad daylight. Not a shot was fired from the great fortress. We walked in a straggling group across at least fifty yards of open sandy ground, making perfect targets for anyone in the fortification less than one hundred yards behind us. There was only one answer, the whole garrison had fled in panic.

Any shred of respect for local fighting qualities evaporated that night. They might be great at long range, where they rarely did any harm anyway, but close quarters – our favourite – they didn't want to know. Irrespective of numbers or reputation they were now definitely on a loser. We knew it, and I expect they did too.

During what was left of the night, and an hour or two of daylight, we hoofed back to Cassino, collecting our three men, then the machine-gun crews on the way. We took each day as it came on operations, never knowing what day of the week it was, or even the date, so it was with some surprise when we reached Cassino that I realised it was Christmas Day. So the Akbat raid went from Christmas Eve to Christmas Day 1958.

During our move back to Cassino, at the first break we had in daylight, I rolled up my trouser legs to check my legs for injury, having remembered feeling things hitting my shins and one knee during the heavier firing. All I found was a slight graze and a few tender spots where there was bruising. All caused, I believe, by chunks of rock thrown up from the ground in front of me. Having had shrapnel before and not even knowing about it at the time, I thought it best to cheek before anything went septic. One of the others found a shiny bit of metal in his upper arm, which he pulled out. It was no problem, not even needing a dressing, just a dab of antiseptic. We all knew how lucky we had been not to have taken any real casualties.

Somewhere in Bahrain or Sharjah, we had an SAS representative looking after our interests at that end of the air-drop scene. He was a humble trooper, or maybe corporal, and – after himself, naturally – he always put his mates first. So it came to pass on Christmas Day we drank his health in lots of lovely whisky, rum, brandy and vodka which, originally destined to grace the bar of an officers' mess, had accidentally found itself with the wrong air-drop label and thereby saved from waste. Sweeny, you played a blinder.

Someone at HQ Persian Gulf also came through on the right line. On the same air-drop we received tinned roast chicken. It worked out one tin between two men. Half a chicken each – happy Christmas.

The Akbat raid had worked out well, but could hardly be claimed as going according to plan as far as 16 Troop were concerned. We had achieved our aim by playing it off the cuff as we went. The rest of the plan had apparently worked like clockwork. 17 Troop achieved their objectives against light resistance. The machine-gunners had come in on cue as required, in spite of the rearranged schedule, and we had no casualties.

One thing had gone seriously wrong. Our 3.5 rockets, which should have had a maximum range of nine hundred yards were pushed to make four hundred yards. Had 16 Troop moved a little further north to find a way across the ravine we would have been hit by our own rockets which were falling two hundred yards short of the target. Also, several rockets had failed to fire from the launcher. Others had failed to explode on impact. Whether it was the altitude or whether sub-standard foreign rockets caused the problem, I never did hear.

Christmas Day we ate and slept; Christmas night we sat around a good fire in a hollow behind Cassino and sampled the good spirit. During the celebrations the opposition regained enough of their composure to give Cassino rather a lot of mortar shells, combined with a steady raking of machine-gun fire. After a few minutes the NFR fired a red signal flare down our side of Cassino. It was the agreed signal showing they needed mortar support fire to calm down the opposition. Our mortar position was about three hundred yards from the hollow where the party was in full swing.

Nothing happened. No mortar support was forthcoming, so a second red flare came sailing down the hillside, then a third and a fourth. Meanwhile the streams of tracer bullets coming over the top of Cassino increased if anything. Being mostly .50-calibre, they went a long way across the sky before burning out. More flares were sent towards the mortar position, until some of them set fire to a lot of grass and scrub on our side of the hill.

The NFR were yelling for our mortar crew to return the fire. Still nothing happened. After several minutes I ran down the hill to the mortar position to see if all was well. One of the crew was sat with his back against the mortar fast asleep. I awoke him and said, 'They want you to do something about those bloody mortars.'

He said, 'What bloody mortars? I haven't heard anything!'

So, I said, 'Well, look behind you, half the bloody mountain's on fire.'

Before I got back to the camp-fire, our mortar was hurling shells back at the opposition who were soon finished for the night. The party, of course, carried on regardless.

Rory Walker was not the greatest piper by a long chalk, but he had his bagpipes there that night, doing his best to terrorise the natives – or us! He was to hit the headlines a few years later when he played his pipes outside the besieged British Embassy in Jakarta, Indonesia. I hoped he had improved by then.

A few days after the Akbat raid we marched down the way we had come up, leaving the NFR company and our mortar crew to hold Cassino while we had a much-needed break. Such was the reputation of the opposition, the NFR would not have stayed on their own, otherwise the mortar crew could have come with us. Shortly after we left the north side, 'A' Squadron, 22 SAS, took over the Cassino position, having been rushed from Malaya to give us a hand.

We returned to Bait-el-Falage for a few days' rest. While there, one or two of us went to see Duke's grave. The only way to the Christian cemetery in those days was by sea. As the cemetery was just up off a beach surrounded by sheer cliffs, George, myself and one or two others hired a boat with an outboard engine (and a driver who knew the cemetery) to make the trip. It being hot and we being thirsty, we took along a crate of beer.

We found the cemetery easily, and, as all the other graves were very old, we found Duke's easily. There were no stones to mark his grave, just a mound of earth, so we gathered some rocks and built a surround. Then put a couple of big rocks at each end. Someone found a stone cup, half buried in the sand, so we put that on top of the grave. Having no flowers we decided we would fill it with beer. It wasn't very big, held about half a pint, and was in the shape of the FA Cup, minus handles. George poured the beer, filling the cup, we thought it was more fitting than flowers – knowing Duke.

We wandered around the cemetery for a few minutes, looking at the names on the ancient headstones, then George gave a yell. 'Who's drunk Duke's beer?' No one had been near the grave since we poured the beer. Looking at that empty cup I felt the hair prickle on the back of my neck. We stared at each other for a few seconds, knowing no one had drunk it, hoping someone would say they had!

I poured more beer into the cup and we left almost right away. The cup was probably carved from porous rock, and the place was like an oven anyway. Nevertheless, it was good to think Duke had enjoyed his beer.

In the few days we were resting, I bought a camera with lots of film to take on the next operation. We fired our weapons quite a bit, zeroing them to our liking. We would have done a lot of swimming but the beach was thick with jellyfish. George said they were harmless and dived in through them – only to come up yelling about ten yards out. They had quite a nasty sting, so we then had the problem of getting George back through them. We eventually got some branches and swept a gap, through which he made a mad dash.

The next time we went to the beach we saw a shark fin cutting through the water about fifty yards out; that was me – and a lot of others – finished with swimming in that area.

Our next operation was on the south side of the Jebel Akhdar Plateau. A place called Tanuf, which gave its name to a monstrous slab of rock, two or three miles long, which appeared to be leaning on the mountain. The Tanuf Slab was where Duke had been killed, but there was no real problem as we now had the measure of the opposition.

Before we went on the next operation we lost our troop sergeant, Herbie, who had a knee injury and, much against everyone's wishes (especially his), had to be rested. He was replaced by Alex, a corporal at the time, who had been on leave in the UK but was able to rejoin the squadron at the end of his leave rather than wait for our return to the UK. Alex was an experienced soldier and well known in the squadron and he had no problem on joining the troop. He had been with 22 SAS since it was formed in 1950.

It was the usual hard, hard climb, loaded down with kit until we reached the top of the slab. It was not as high as Cassino by a long way, but the opposition were very touchy about its occupation as they knew it was closer to the main part of the plateau and made a good base from which to chip our way upward and forward. Right at the top of the slab we organised a good defensive position, which was promptly mortared from a position high up and far away, beyond the reach of our 3-inch mortars. Our *sangar* walls were strengthened and grew upwards rapidly.

The day after that they gave us a short pasting with a .50-calibre machine-gun. Our .30-calibre machine-guns couldn't compete. The *sangar* walls were all thickened some more before nightfall. No one was hurt by all this show of fire-power. We knew someone would get hurt when we closed with the opposition though. We were also very confident it would not be us!

As far as I remember, the squadron commander went up the Tanuf Slab with us, then had to go down again for one of those conferences which always seem to bug good field commanders. On his return to the squadron position he brought some mail. After the mail was doled out he called me over. He was still showing signs of the hard climb back up the Slab. (No helicopters for mere squadron commanders in those days!)

His first words took me by surprise. 'Why haven't you written to your wife?' I told him I had written at least two letters while we were at Bait-el-Falage, soon after Christmas. The Major looked decidedly suspicious. 'She hadn't had a letter since you left Malaya, so you'll bloody well write now – and give me the letter when you've finished!' With that he handed me a writing pad, envelope and stamp.

I don't know what had happened to the several letters I had sent Ann from Oman, but there was no point in arguing over it. I wrote a letter, handed it to the squadron commander, who said it would be sent out at the first opportunity. I retired to my *sangar* feeling a bloody fool but anxious over Ann's worries if she had not received any mail since I left.

Later, I found the situation had been made worse by one of Ann's friends getting one or two letters every week from her husband. The fact that he was a signaller in base all the time was, of course, lost on the two wives.

During the hours of daylight we were on constant watch with binoculars. Every movement by the enemy was carefully watched, very often when they thought they were too far away to be seen. On one of these occasions we watched several of the enemy carrying something we couldn't identify into a cave about three thousand yards away. It was obvious they were not doing something for the benefit of our health so, as we had just been joined by a RAF Forward Air Controller it was a good chance to try him out. He could speak to the jet pilots when they were overhead and direct their attack.

The RAF man was a pilot from the Venom Squadron, so had the great advantage of knowing the pilots he was directing. He was also an ex-member of the Rhodesian SAS (our 'C' Squadron), so fitted well with us. His head was shaved down to the wood and his code name was 'Goldilocks'. Very fitting.

The Venoms were sent for. In the meantime our Commanding Officer, Lt-Colonel Deane-Drummond, who had arrived in the helicopter to have a look at the position, was about to leave. The jets were in contact with Goldilocks just as the chopper started to take off.

At that moment we found out what the opposition had been carrying into that cave. A .50-calibre opened up, trying to hit the chopper as it lifted off. The chopper dropped down the mountainside unharmed and out of the line of fire. So the .50 laced the position with long, stringy bursts for a minute or so.

The heavy rounds smashed into *sangar* walls and seemed to explode rocks all over the place. Ricochets, pieces of bullet and lumps of rock screeched and whined everywhere. Several of us pinpointed the source of all this uproar as the cave, which was way out of effective range for any weapons we had with us. The location of the enemy was made easier for us by the smoke around the .50, which caused some wit to remark how handy it was that they had used camel fat to oil the weapon.

In spite of the ferocity of the enemy fire, we were not much bothered as our defences had already been well strengthened and even a .50 would not make much impression.

Goldilocks now came into his own. His mates were circling way up out of sight to the south, and he got busy with the target description. The flight leader came hurtling over us and fired a marker burst from his 20-mm cannon, which struck way above the cave. Goldilocks corrected from that burst, and the second jet put his shells around the cave. The first jet came back and fired one rocket which struck about ten feet above the cave. The second jet came screaming over us and placed his rocket straight in the cave mouth. It was great to watch. Those people must have thought the return fire was fantastically fast and effective. They would never know it was on its way before they had even fired.

A day or two after that incident we moved off the Tanuf Slab towards the main plateau. We only moved a couple of thousand yards. The main purpose of the move was to keep the opposition on their toes and also to check out the ground and any natural obstacles between us and the villages on the plateau, which were the enemy's main bases and supply areas.

We reached a very rough, jagged rock area in the early hours and settled into defensive positions. Soon after settling down, before daylight, the opposition fired a couple of ranging shells with an 81-mm mortar. These fell about three hundred yards south of us. A few minutes later they began firing again, at a slow rate, putting down twenty-five shells in the area of their ranging shots. Realising they had made another cock-up, I went to sleep before they finished, something I thought I could never do with mortar shells swishing overhead.

On this operation Rory Walker, who had led the troop so well and was very well liked by everyone, had been put on another job: co-ordinating air supply from the sharp end.

From our new forward base it was planned to do a reconnaissance on to the plateau. The squadron commander, John Watts, picked three of us from 16 Troop to accompany him on a patrol through the enemy defences. 17 Troop and the rest of 16 Troop were to provide step-up for the first part of the patrol and give us cover if we had problems getting through.

About one thousand yards forward from our start line, just before parting with our step-up group, we had a head-on contact with a strong enemy force. A bright moon, which I had found disquieting with the prospect of climbing towards it through enemy positions, had, in fact, the opposite effect when meeting the enemy in the open. They had their backs to the moon, therefore when we met they were easily seen as black shadows. Perhaps our leading scouts opened fire too soon to obtain maximum effect, but at least we had the drop on them and they withdrew.

We had been moving parallel to the enemy defence line, at about four hundred yards' range, and now came under fire from their forward positions. Looking at all those muzzle flashes, I realised we would have had some problems getting through undetected, even in a small group. Although there were obviously a lot of people firing at us, there was comparatively very little fire actually coming our way. It was more like being a spectator at someone else's battle, with the odd few shots accidentally crackling around us. The rocks in the area magnified the sound, gave an uproar of echoes and causing ricochets galore, added that savage screech to the overall row.

A couple of our Bren-gunners and a rocket-launcher crew returned the fire on enemy machine-gun positions. I don't think any rifles were fired by our people after the initial contact. Sharing a large rock with a friend, we peered round our respective corners. I estimated the nearest enemy position to be at least three hundred yards from us. He obviously thought the same. 'Too far to bother,' was his verdict.

After a minute or so the enemy fire slackened enough for us to sit on our rock in comfort, awaiting a decision for our next move. It was not long in coming. The enemy were pretty thick on the ground in that sector, so John Watts decided to call off our patrol for the time being, as the enemy were now all stirred up and would be extra alert. This was very likely another of those cases where the squadron commander's decision was heavily influenced by our top priority: no casualties.

The patrol move through enemy lines at that stage was not crucial to the operation. Had things been different, it's likely a dummy frontal attack on an enemy, who was obviously jittery and couldn't hit a barn at night, would have caused ample confusion for us to have slipped through with little risk. Personally, I would rather have gone through the defences with a gun in one hand and a grenade in the other, so to speak, than try to tiptoe through their kitchen.

The group commenced withdrawal to more friendly territory. All firing on our part ceased, as it would show our movement. We checked for casualties and found one man was missing.

There was no confusion at the time, although we were still not actually being fired upon, but we were in an area where a fair amount of enemy rifle and machine-gun ammunition was banging and screeching around, or crackling overhead. While the main party returned to our forward base, sweeping the area for the missing man as they went, four or five of us began sweeping the area to our furthest point of movement that night.

Searching for one man – who could have been wounded, unconscious or dead – in an area covered in man-sized rocks and at night, is a rather hopeless task. The moon and stars came to our aid on this occasion but this made me, at least, feel rather exposed to the opposition.

We formed a line and moved slowly over the areas of possibility, feeling in the darker shadows, staring hard at suspect shapes. The opposition settled down to the occasional raking of the area with machine-gun fire which, although having its nuisance value, was also very handy as it indicated there were none of their people in the area. After half an hour or so of fruitless search Alex called us off and we commenced our move back to the forward base, still in our search line.

About halfway back we met two men coming out to meet us. The missing man had been found. It had all been a mistake, no one had really been missing. A waste of time, perhaps, but it had brought out the problems of searching ground under those conditions – and the importance of being one hundred per cent sure of where everyone was. There was no one to blame, it was just one of those mistakes which can happen in a noisy, dark situation.

The squadron stayed in that forward position for a couple of days but, as far as I know, or remember, we were not bothered by enemy activity. There may have been the usual hopeless, long-range sniping but nothing of note. A day or two later we all withdrew from the Jebel, marching back down to Tanuf for a few hours' break. Something was in the air. We would find out soon enough.

11 AKHDAR – FINAL ASSAULT

Our next trick was rather complicated. 'A' Squadron joined us from the north side. Both squadrons were to be used in an all-out assault to finish the job. Perhaps I missed something along the way, but I don't remember knowing the name of the place or area where we were going to do this final assault. I knew it was several miles southeast of Tanuf and we were fairly well briefed on the route to be taken up the Jebel.

A couple of 'D' Squadron people had done a sneaky recce of the lower part of the route and had studied the climb in some detail from out on the desert, but they couldn't take any risk of alerting the enemy to our interest in the area. So there was a lot of problems which could not be solved in advance. We studied air photos of the route but none I saw showed the view from directly above.

I heard later that the local donkey wallopers, who were being used to help with supplies on the south side were sworn to secrecy, on pain of death, not to breathe a word about where they were to take the next lot of supplies. We knew that by telling them it ensured the enemy would know all about our plans. But, of course, the donkey wallopers had been given the wrong location.

We spent a day checking and zero-firing weapons and checking all our kit: ammunition, grenades, rockets, rations, everything down to the smallest detail. 16 Troop, under Alex's direction, ran through all our contact drills and immediate action (IA) drills.

Our main load was ammunition. I remember having two 3.5 rockets, four ninety (Energa) grenades (now issued with correct cartridges). Eight No. 36 grenades, six No. 80 (white phosphorous) grenades. Five twenty-round magazines of rifle ammunition, plus one hundred rounds in bandoliers and one two-hundred-and-fifty-round box of .30-calibre machine-gun ammunition.

Hanging on the back of a truck, I found one of those spring balance weighing hooks and, just for curiosity, weighed my kit. My Bergen rucksack, loaded and ready to go, weighed ninety-eight pounds while my belt weighed twenty-two pounds – a total of a hundred and twenty pounds. I don't remember if that included my rifle. It probably didn't, as I don't remember putting it with either belt or Bergen. Everyone had similar loads to carry. We wondered what the night would bring in the way of 'uphill'. Any opposition we met would be welcomed as an excuse to unload some weight.

Final checks were made and we climbed aboard trucks. There were about fifteen or twenty vehicles in the convoy, which headed north from Tanuf, on the rough dirt road and tracks along the base of the Jebel. The trucks were well sandbagged as protection against mines. 16 Troop were in the leading truck. It was still broad daylight and we were clearly visible to the watching eyes on the Jebel. At the back of the convoy was the water tanker. In the passenger seat of the tanker sat Bill, an SAS medic corporal who had already survived being blown up twice on mines, once while in an ammunition truck half-full of mortar shells. So Bill travelled at the back of the convoy – the safest place in respect of mines. Just before dusk we saw a great mushroom of smoke and dust go up at the rear of the convoy. A few seconds later the thunderous bang of an anti-tank mine reached our ears.

The convoy stopped and John Watts rushed back to check the problem. A few minutes later he came back and said, 'You're not going to believe this.' Poor old Bill was only badly shaken having gone out through the observation hatch of the cab on the water tanker.

How could they always pick on Bill? The whole convoy had trampled over that mine, but it waited especially for Bill's water tanker. There was no real mystery about the mine. It was likely a Ratchet mine, which allows so many vehicles to pass over it before detonating. Very handy for mining busy tracks and allowing the minelayer to be long gone before anything happens. Nevertheless, Bill's attraction for mines was phenomenal.

After dusk the convoy turned around and, without lights, rushed southwards through Tanuf and on into the desert, parallel with the Jebel. At the right spot we turned off the track and headed straight for the foot of the Jebel. Reaching the steep slope up from the desert, the trucks stopped, we piled out, dragged our heavy loads on to our backs and strung out in single file to make the climb of our lives.

There was no real track at the point we went up. We had seen air photos of the route, taken by a Gloster Meteor recce jet. The only photos I had seen were what are called 'nose-face' photos, taken by the nose cameras of an aircraft flying straight at the area to be photographed. So we had at least seen a view of the route, as seen from about two thousand feet above the desert as the aircraft flew towards the Jebel.

We knew we had to find our way more or less straight up the side of the Jebel, a climb of roughly eight thousand feet. The actual climb involved more than was apparent at first sight, as the air photos had not shown dead ground. Somewhere about halfway up we had to climb down about two hundred feet, almost sheer. Ropes had to be used, as climbing down in almost total darkness can be tricky, we couldn't see the bottom, or what sheer drops were below us.

Going up was not so bad, but the weight we were carrying began to tell about two thousand feet from the top. Men began to collapse in their tracks. I saw two fall flat on their faces, unconscious before they hit the ground. Those of us who could kept going stopped for nothing. I personally could think of nothing except those machine-guns at the top – and the ball they would have when daylight came if we didn't get to them first.

Men who fell got to their feet again and staggered on. There is no greater spur to keep going than the thought of a gut full of .50 calibre if you don't. When I reached the top it was a false crest. Another sheer drop lay before us, but the main plateau was a relatively easy climb from the bottom of it. We could see the top of the plateau against the sky. It appeared to be not more than five hundred or six hundred feet higher than our false crest, although it was difficult to judge in the dark as we were not sure how far away it was.

The cliff was only forty or fifty feet but the first five or six of us to arrive hadn't a rope between us. The last two thousand feet had been a matter of every man for himself, heading for the lowest point on the skyline ahead, as instructed.

There were no officers present with the first few at the top, but the decision was to consolidate at that false crest until more men arrived. It was obvious to everyone that the false crest had to be the enemy defence line, as it had a commanding view of the slopes we had climbed, as well as the final slope of the plateau.

Our surroundings were, of course, very dim outlines against the night sky. Distances were difficult to assess. We worked to build *sangars* in almost total darkness and as silently as possible. Picking up loose rocks from all around I suddenly noticed I had got shit on my hands. I realised I was using rocks which had been used as covers by someone cat crapping. I spread the word and almost immediately someone found a .50-calibre machine-gun. It was on an anti-aircraft mounting, loaded and ready to go. There was no sign of the crew, so we carried on preparing our defences as more men arrived in ones, twos and threes.

Where there's one gun, there are usually more not far away. As our strength increased, men went silently searching among the great rocks and jagged outcrops all around. Someone with a rope had arrived, so we fixed the rope to get down the cliff and, as all of 'D' Squadron were now ready to go, 17, 18 and 19 Troops went on ahead leaving us to give covering fire if required and to solve the problem of the machine-gun crew and any other problems in the enemy defence line.

Daylight started to creep across the landscape and we soon found the caves. They were on the cliff side of the ridge which formed the false crest, facing the main Jebel. Two of 16 Troop, George and Alex, went to sort out the lower cave. My patrol commander, Miley, and I approached the caves nearer the gun position. We waited for the daylight to strengthen, then moved in. The first two caves, not being deep, were easily checked. The third went well into the rock face. It was my turn to make the entry so I went in fast, prepared (I hoped) for anything. Nothing in the outer chamber, but the roof was low at the back and it looked like an inner chamber might yield something. I signalled Miley to go through as he was much smaller and would be able to cope better in the confined space. He did a fast check and found nothing except grass which looked like it was used as a bed.

As we came out of the cave a few shots rang out, then a grenade blew, so we knew Alex and George had found the gun crew. Alex had been climbing down to the cave and had almost reached the ledge in front of it, unable to use his rifle, when one of the enemy came out to stretch himself. Alex has always claimed George waited to the last second before firing – just to give him a hard time!

With that one out of the way Alex pulled himself on to the ledge and put a grenade into the cave. That took care of the other occupants and the problem was solved.

That we had caught the opposition asleep, with no sentries or lookouts posted, no forward listening posts and – as we found later – pretty thin on the ground, was more proof of bad leadership. Presumably the main rebel force had been bluffed into the wrong defensive area by our tales to the donkey wallopers and deceitful daylight moves by the convoy of troop carriers, but it would have to be a very naive soldier who didn't have a fifty per cent stand-to during the hours of darkness when the enemy was obviously up to some new skulduggery.

Our move up the almost sheer side of the Jebel had been much too noisy for our liking. We had sacrificed silence for speed. Although we had kept as quiet as possible there is no way a man can climb over loose rocks and shale, heave himself up rock slabs (and fall down a few) in almost total darkness with a heavy load and, at the same time, be both fast and quiet.

Alerted by the row Alex and George had made, another machine-gun crew got busy on us from about five hundred to six hundred yards away. Luckily it was only a light machine-gun and its fire was almost ineffective. Nevertheless, being subjected to a hail of machine-gun fire in that terrain is a very noisy and off-putting experience. The bullets crackled overhead, banged ferociously against rocks and ricocheted noisily all over the place.

After the first few seconds some of our people were standing in the open, looking for the enemy gun position through binoculars. This was very off-putting to the gunner, who obviously could not see where all his fire was going. He started making wild corrections which took most of the heat off us. We very quickly located him above and to the east of our position – near the top of a castle-like, sheer-sided pile of rock. None of us had, as yet, fired a shot in that direction.

A half-belt of .30-calibre return fire by a couple of 16 Troop more or less cleared that problem, so that the enemy were reduced to the occasional nervous burst of fire when they thought no one was looking (that's how it seemed to us anyway). There were, I believe, a few other machine-gun positions put out of action by members of 'D' Squadron that night and early morning, but I saw nothing of them.

Daylight had shown us the edge of the main plateau was about eight hundred yards away and about six hundred to seven hundred feet higher than our position. There was a flattish wadi between the foot of our cliff and where the last climb started, about two hundred yards away. From there a track showed clearly, climbing away up over the shoulder of the comparatively rolling hillside of the plateau.

Rory Walker had arrived, carrying a backpack which looked like a small mountain. He decided to press on in company with his old troop. But his load was just too much for him to keep up as we belted off up on to the main plateau. Seeing the situation, I volunteered to stay back to keep an eye on him rather than see him left on his own, looking rather small and alone in the vastness of enemy territory. He caught up with me and we went on together awhile. Before clearing the last crest we stopped for a breather.

Looking back down to our last position, we could see 'A' Squadron had just started down the rope. What happened next was really comical. Two of their people were at the bottom of the rope, one was halfway down and another was just starting down, when that machine-gun fired another nervous, ineffective burst. One of the men at the bottom of the rope tried to go back up, the one at the middle tried to go back up, the one at the top slid down and demolished the lot at the bottom.

Then they started shooting in various directions. Rory and I looked at each other, then both slid around the rocks we were leaning against, to put them between us and 'A' Squadron.

Goldilocks must have been around somewhere as, soon after that, the Venoms appeared and blasted the machine-gun into eternal silence. The

Venoms then attacked other positions along the enemy defence line, east and west of where we crossed it. Whether they were just hitting likely places or were being directed to enemy positions I don't know. It is most likely they were being directed to enemy positions found by 'A' Squadron as I had never seen them waste ammunition before. It was the last time I remember seeing the Venoms on that operation. The tearing roar of the rockets, their exploding crash, the crackling and crash of cannon fire and the occasional, rather pathetic sound of answering machine-gun fire along with the scream of the jets spoiled the calm silence of a beautiful morning, peculiar to high, wild places.

Rory and I plodded on and soon found 'D' Squadron, who were getting patrols organised to look for a fight. So far the expected battle had been a big fizzle. Rory did his thing with the air-drop markers and smoke grenades. The RAF came in right on the button with the first air-drop. I have always felt sure those several supply drops that morning had more to do with winning that campaign than any other single factor.

The enemy listened to Radio Cairo, who claimed every other week that another British airborne brigade had been wiped out in Northern Oman. I can imagine the enemy lapping up all this heroic crap but knowing in their hearts they had not wiped out anything. Then the moment of truth when the sky over the plateau was thick with parachutes, load after load. To an enemy unfamiliar with airborne operations, it would have been obvious to all who weren't really close to the air-drop that those airborne brigades Radio Cairo thought they had wiped out were now arriving *en masse*! So they all ran away – and lived happily ever after – much to our disgust after all the effort we had put in to getting close to them.

Of course, at the time of the supply drop, we didn't know all that, and were expecting a real battle to develop any minute. The crew of one Hastings aircraft complained bitterly they were being shot at, so off we went on patrols to find the culprits.

George, myself and a Bren-gunner found a long cliff full of caves – where we thought the .50-calibre had fired from. We spent a hairy half-hour going from one cave to the next, checking them out. Checking a cave is a particularly tricky business. Anyone in the cave has the obvious advantage of waiting comfortably in the dark watching for the visitor to show up clearly in the entrance. There is no way to enter a cave without making a beautiful target of yourself, except by using explosives or smoke, or waiting for a very dark night.

Then there is the problem of adjusting one's eyes to the darkness. By the time your eyes are of any real use the opposition has had about two minutes in which to fill you full of nasty little holes. We didn't have the time to wait for a dark night.

There were at least thirty cave entrances to check, along a cliff about three hundred yards long. Explosives and smoke were a possibility, but we still had a war to fight and would likely need our grenades much more urgently before it was finished. So, as usual, we leaned on hard experience, fast reactions, teamwork and good old Lady Luck. Experience has shown that a fast, awkward moving target is very difficult to hit, especially when it arrives suddenly, even when half expected.

The gamble on our side was (partly) that the first enemy shots would not hit us, whereas our fast reaction to any gun flash would – if not disabling the enemy – at least put us on more equal terms for the vital seconds which decided who lived and who didn't.

George and I took it in turn. That was the other part of the gamble; Russian roulette with a different twist. Where there were defensive walls just inside the cave entrance we would throw a grenade-size rock as we went in.

The Bren-gunner's job was vital and tricky. Ideally, he should be in a position to fire into the cave if the first man was hit so that the second man could pull his wounded comrade clear, or solve the problem before the enemy gained the initiative. The Bren-gunner was also responsible for the suppression of any interference from other quarters while his mates were busy. On this occasion our Bren-gunner could not move too far forward without being at risk from other uncleared caves so he used the same tactics as a winger on the football field, positioning himself to bounce them in off the far post. In fact, with big caves that is by far the best method when your own men are in trouble at the entrance. The machine-gun fire hitting the side wall sends a hail of ricochets and rock fragments into the cave and at the same time creates a pall of dust and smoke to cover the recovery of wounded and the follow-up grenading.

The caves we found that day varied considerably: some were very big at the entrance (ten to fifteen feet high and ten to twenty feet wide); others had a small entrance (four or five feet square) but were quite big inside. Several had been recently used, some had not been entered for years.

The first thing we checked for was recent tracks, human footprints in the dust. Some were so well used there was no dust in the entrance. Most of these had a defensive *sangar* at the entrance and another a few yards inside the cave. Very sweat-making and hard on the adrenalin pump!

We would wait a few seconds, listening for any tell-tale sound, sniffing for smoke or smell of scent; a final check that the other two were in position, then rush through the opening or dive over the wall or maybe roll across the floor towards a handy-looking bit of cover, depending on the set-up and what our nerves told us. There was another worry in our minds: we could barge in on a perfectly innocent shepherd family hiding from the war. We had discussed this possibility briefly before starting and decided we would cross that bridge when we reached it, but in any case no chances were to be taken. 'If in doubt – lash out!'

We found nothing, and returned to the squadron position to find no one else had found anything either. About this stage in the game we heard 'A' Squadron had taken casualties following us up the mountain. A sniper had hit a grenade or rocket on a man's belt, killing him and the man behind him and seriously wounding another man. We were dismayed to think there had to be casualties in the regiment when we had achieved so much without serious resistance.

I don't know if it was that day or the next that 'D' Squadron again moved out to look for the enemy. The first village we came to was named Habib. The houses were mostly very small, with only two or three small rooms, all at ground level. They were built with mud and rock walls. Some had been patched with corrugated iron or tin sheets. I don't remember much about the roofs but I'm pretty sure they were mostly thatched or corrugated tin and almost flat. The buildings were crowded together in a small area about the size of a football field or smaller, as if huddling together for protection from the savage, jagged rocks all around. Their windows and doors were all very small, most windows being little more than slits in the walls. Very few had any glass in them, not because of the war, as there was no war damage that I could see, but because no glass had ever been put in them. Some had rough wooden shutters, others were stuffed with sacking or had nothing.

A few men were sent to have a quick check around the village but not many buildings were entered. We had no wish to enter people's homes if there was no reason, no threat to us. Most of the doors were closed but not locked, almost all were held shut by a rock or piece of wood leaning against the outside, as if to say, 'There's no one inside.' A few houses did not appear to have a door, just an empty black doorway. I personally saw none of our people enter a house. An empty house – or an empty village – has a certain feel about it. I had no tingles, no quickening of the pulse when I stepped into the village street. My soldier's instincts told me there was no danger. Unexplainable. I didn't even think about it at the time but, long afterwards, thinking back, strange.

I crouched in a couple of doorways to check the interiors, not the way to operate if there is the slightest suspicion of danger. The reason for crouching was to let light enter through the doorway, which was the main source of interior light. That way I didn't have to enter the building and step clear of the doorway. Afterwards it seemed all the more remarkable when remembering it was the first village we had entered in enemy territory. It was completely deserted except for a big old brahma bull we found locked in a shed. He looked a bit hungry so we let him out to forage.

We had received twenty-four-hour pack rations on the air-drop. These were in a cardboard box which contained three separate meal packs, each marked with its intended place in the day: 'Breakfast', 'Snack' and 'Main Meal'. We were sitting well spread along a narrow street in Habib, having a meal break when the bull came wandering up and picked up an empty snack pack. I took a photo of it (since lost) with the word 'Snack' showing clearly like a speech balloon in a cartoon. It seemed to like hardtack biscuits better, though.

From Habib we moved on to Saiq, the main village on the Jebel and headquarters of the enemy. The move from the air-drop site to Habib was done in column, due to the terrain, but soon after leaving Habib we came to a wide-open area covered in loose rocks and very low scrub. Expecting to hit heavy enemy resistance any minute, the squadron spread out into a wide skirmish line with everyone ready for the first signs of trouble. I don't remember seeing anything of 'A' Squadron after leaving the airdrop site, so I don't know where they went. As far as I know, the forty-odd gun hands of 'D' Squadron were on their own.

I do remember someone saying, 'If they're ever going to hit us, it's got to be in the next thousand yards.' The further we went out across that comparatively open area, the more I felt we were walking towards a helluva barney. Through habit, I looked ahead for my next bit of cover, noting its position, then gazing all around – with a quick glance over my shoulder for good measure – reaching the rock or small depression I'd noted for cover, looking ahead for the next and so on. We were all doing the same, looking for possible gun positions, any sign of movement, a glint of sun on metal. We looked for low mounds or slight depressions in the ground around us which would favour fast safe movement on all fours – or our bellies – if the need arose. It's too damn late to start looking once the good old brown has hit the fan. Our ears strained for the first crackle and screech of machine-gun fire, or the muffled bark of a mortar.

Nothing happened. Where were the five hundred well-trained rebel troops? Where were the two thousand or more savage mountain tribesmen? Surely someone would have to have a go at just forty of us?

Dirty, unshaven, our sweaty hands holding rifles or machine-guns, every now and then shoving a thumb under the pack straps to ease the weight and prevent an arm from going numb, we plodded on. I think we all subconsciously wiped from our minds the maximum possible odds we could come up against. Allowing for a couple of thousand away on other business we could still be faced by odds of over sixty to one, but we had decided long ago the odds were never likely to be much more than ten to one, so there would be no problem.

No odds materialised. No one interfered with our walk in the sunshine. Nothing happened.

I think I can claim to be the first invading soldier to see Saiq from the ground. I happened to be left winger of 'D' Squadron when we approached the village, and stopped the squadron when I reported. John Watts came over to where I stood on the edge of the wadi cliff, about three hundred yards from the village. He said, 'That's Saiq, for sure.' Then he issued orders for the final approach and securing of the village.

Saiq, too, had been completely abandoned. Not a living thing in the place. I could 'feel' eyes though, from the rough and rugged ground surrounding us. The village was bigger than Habib, boasting several two-storey buildings. It had been hit by air attack during the last few months but not many buildings were badly damaged. The damage could all have been done by three or four rockets and a few dozen cannon shells. It was known that the rebel leaders had made it their headquarters, supposedly having bomb-proof caves somewhere north of the village. These had been the target for the heavy bombers which had, as far as I could see, not touched the village.

I was not involved in the house-to-house check through the village so did not see it all at close quarters, being one of two men sent to cover one of the routes out of the village. We saw no one, just a few chickens. Once the place had been checked we set up base (threw down our packs and got a radio working) on the south side of the wadi, opposite the village. The wadi was about fifty yards wide at that point and only about twenty feet deep. We set up a loose defensive area with half-hearted *sangars*. There was no real need to entrench ourselves. If an enemy turned up, we would do the attacking, we would have to in order to keep the initiative. In such small numbers and against heavy odds, we could not turn ourselves into a compact target to be mortared and machine-gunned. There was natural cover everywhere around us, so we used it.

Patrols of two and three men were sent in all directions to check for signs of the enemy. Booby traps were a possibility, as were ambushes and mines, or combinations of all three. Miley and I checked a deep wadi just below the village. It looked well used and we soon found a large cave. Going through our familiar caving routine we were soon inside, and getting our eyes accustomed to the darkness discovered we had hit the jackpot. The cave was full of weapons.

I don't remember the quantity of all the weapons, but I do remember counting twelve Bren guns, each with its box of magazines. There were a lot of .50-calibre machine-guns, one or two mortars and a lot of ammunition for everything. In the middle of the main chamber there was a pile of blankets, which suddenly moved. We were both on our nerve ends, moving very carefully, half expecting a hoard like this to be booby-trapped. When the blankets moved we both dropped and aimed at the point of movement.

A little whimpering sound came from the blankets. Miley stepped forward and quickly snatched the top blankets away. Imagine our complete amazement to find ourselves staring into the tearful eyes of a two-year-old girl. I quickly went past and checked the rest of the cave. Miley just stared at the kid. I don't think he realised his rifle muzzle was about six inches from her face. He was waiting for me to finish the check. When I found nothing he moved the rifle and clicked the safety on.

We got out of the cave, Miley carrying the child. A supply drop was just coming in by the village. The child was obviously terrified of the aircraft. We calmed her down and gave her some sweets from the rations we carried. By the time we had climbed up out of the wadi she had, as children do, accepted the new situation, laughing and pointing at the coloured parachutes floating down nearby.

Up on the flat ground again we met Bob, a corporal in 18 Troop. We had a quick conference and it was decided the best thing for all concerned was for the child to be returned to her mother as soon as possible. Bob said he thought he had seen some civilians about six hundred yards down the slope, south of the village. We went about two hundred yards in that direction, Miley carrying the child.

At the top of the slope we stopped. It was decided that Bob and Miley would leave their rifles with me and take the child to the bottom of the slope where there was rough ground, huge rocks and a few scrub trees. I complained there was little I could do to cover them at that range but they thought it was a risk which had to be taken.

When Bob and Miley were about two hundred yards from me, a woman ran from behind a big rock, then ran back out of sight again. Bob and Miley went another hundred yards towards where the woman had shown herself, then stopped. Bob lifted the child up in the air, then put her down gently on a big flat rock. They then turned and walked slowly back up the slope. When they were halfway back to me a woman ran from behind the big rock, collected the child and bolted back again.

As Bob and Miley rejoined me, we could see several people now standing beyond the rock, looking towards us. We waved to them, then walked back towards Saiq. Ten minutes after we arrived back at the squadron position, a dozen people, women and old men, came to us from down the slope. They were welcomed, offered water and food.

Within twenty minutes forty people had come up the slope to us, and looking very relieved it was all over. They had had a hard time with the ever present threat of air attack. The village had been strafed several times over the past months.

It seems no one learned much from Hermann Goering's experience. Governments and top brass alike still think people can be subdued by air power. The facts of life are much different. People are more likely to be alienated by air attacks. The only way to take ground is to put upon it a man in the right uniform, with a gun in his hand. Bombs cannot discriminate – as a soldier can – between an enemy soldier and a little girl.

Some of the civilians went out into the countryside to bring in others who had been too frightened to return to their homes. By evening the people were allowed to return to the village. We were beginning to realise there was to be no final showdown. The enemy had been defeated with a minimum of bloodshed.

Just north of the village, about four hundred yards up the wadi, was a big cave with a small entrance. It was the bomb-proof hideaway of one of the rebel leaders. Our patrol was given the task of searching it and keeping everyone well away from it. The cave contained a lot of paper records which, we heard afterwards, incriminated a lot of people who had been playing both ends against the middle. It also contained weapons, ammunition and money (Austrian 'Maria Theresa' dollars) as well as rich household effects.

We were rather amused at the way the locals tried to get our permission to enter the cave. First an old man came and pleaded almost tearfully for us to let him enter. Then another old man who walked imperiously with a stick. He glared at us and tried to push past us, demanding to be allowed into the cave. Finally, he flourished his stick in our faces, ranting and raving, almost frothing at the mouth. Then stamping off back to the village with many backward glances and cutting motions with the stick. I think he could easily have bitten a nail in half.

Our next callers were two typical 'mothers-in-law'. Smiling at first, but soon working up to screaming obvious abuse and even trying physically to push into the cave. They had obviously twigged we were harmless to civilians who didn't carry guns. Their shrill ranting could easily have changed my mind though!

The last try was the best. Just before dusk three people appeared, approaching up the wadi. The larger of the three figures, a woman, sat on a rock about one hundred yards from us. After a short conference, which we couldn't hear, the other two came to us and signalled they wanted into the cave. They were two young teenage girls, reasonably smooth, all big eyes and flashing teeth. We had to laugh at their antics to get past us, but they were no Mata Haris, just two good-looking kids who had been elected to crack the problem.

When they got no change, they giggled at one another and looked back towards the black-clad lady on the rock, who screamed something at them which obviously meant 'keep trying'. When they still failed to get past us they sobered up quite a bit and went back to Mum, or whoever she was, where they got a good ticking off by the sound of it, then the whole party withdrew slowly towards the village in the gathering dusk.

There were more tries the next day with at least one old man and two or three women who came alone at different times. No one cracked it while we were there, but full marks for trying.

A few days after taking the villages, 'D' Squadron marched down off the Jebel. We had been told as soon as the fighting was finished we would be sent back to the UK. The operations in Malaya were now over, and the regiment was to be based in the UK for the first time. But the best-laid plans of mice and men . . . We were fated to remain in Oman a further five or six weeks. The squadron did what was called 'flag marches'. A typical bloody waste of energy, plodding to and fro over the Jebel in a so-called show of strength. It pleased the brass no doubt, but the troggery, who continually pulled the top brass's chestnuts out of the fire by their sheer guts and determination are not at all keen on flogging up empty mountains – especially ones they have just emptied. There would have been a lot more people impressed if the vast army of brass had waddled its way up and down a few times instead of doing it all by helicopter. They might just have appreciated what it was like to nip up eight thousand feet on a dark night in about six hours with over one hundred pounds of kit. Don't know what I'm bitching about anyway, I got myself a stomach infection and never did the flag marches. Instead, I was employed at Bait-el-Falage, helping with the trucks, issuing petrol, oil, water and doing the paperwork. A lousy job, in a lousy place – but still better than flogging up empty mountains!

A British cemetery was organised at our camp, to take the bodies of our men killed on operations. While it was being built someone pinched the truck I was using to shift petrol and left it parked in the cemetery. I collected the truck and, as I approached the gateway coming out, a child ran through the opening in front of me causing me to swerve and hit the gatepost. The gatepost was made of concrete, about two foot thick, but the impact toppled it, the truck was hardly marked.

The 'A' Squadron set-up, which was running the base, decided I had to be charged over the cemetery gate affair. So, next day, I was on the mat. No one had asked me anything about the accident so I had no chance to explain anything. 'A' Squadron sergeant-major marched me in with great glee. The charge was read. It was a cock-up! I was charged with: 'Taking an army vehicle without permission.' After checking with the MT Office and finding I had permission to use the truck, they were all at sea as to what to do and eventually told me to get out of the office. That was the last I heard of it.

By the time we left Muscat and Oman we were all damn glad to go. Not because of the operations, but because of the stupidities which followed them.

One thing which may puzzle the reader who has any knowledge of military formations and minor tactics is the fact that we often operated in twos and threes, although the basic patrol is based on four men. In fact, the four-man patrol has been proved many times to be the ideal group for most SAS operations. In Northern Oman we were organised in four-man patrols, and those patrols, as far as is I know, remained the same through the operation. But the SAS is such that through training and operational experience almost any two or three men can work extremely well together, especially men from the same squadron. So, when you are up against long odds, the tendency is to use smaller groups (of two or three) on jobs they can cope with, so as to give a wider spread, cover more ground, do more jobs. Of the men named in the Oman operation only Miley and I were in the same patrol.

In the situation where I was left lagging behind with the village idiot and his donkey, for instance, other members of 16 Troop kept an eye on our progress and, often, on looking up to the skyline above, I would see the welcome sight of one of them watching to check there was no problem. They would give a wave of a hand to say all was well with my surroundings, which they could obviously see much better than I could. An answering wave and they would disappear, then someone would appear much further up the mountain a few minutes later. This was not prearranged, just commonsense.

Most of 'D' Squadron people, including the squadron commander, had occasion to move around on their own during the operation, both in daylight and in darkness. Sometimes there was a calculated risk, usually the risk was negligible. The where, when and how is a matter of professional know-how.

The Jebel Akhdar operation was of great importance to the survival of the SAS. At that time the Regiment was not a permanent fixture in the British army. It was not part of the army 'Establishment' and could be disbanded at any time.

The SAS regiments which were formed during the Second World War, and which proved to be very effective in their role behind enemy lines, had been disbanded very quickly after the end of the war. 22 SAS had been formed specifically for the Malayan Emergency and would, in all probability, have been disbanded when that situation was resolved. As it was, our 'B' Squadron was disbanded soon after their return to the UK in 1959, leaving us with just two Sabre Squadrons. The Northern Oman operation proved the SAS could operate just as effectively outside the jungle, in totally different terrain.

So, the powers-that-be had two main points to consider. Sending a very small force of well-trained men to solve a problem was far cheaper than using conventional forces. In the case of Jebel Akhdar, a force of at least brigade strength (three to four thousand men) had been considered reasonable for a successful conclusion. The other point was just as important. We were so few that no one knew of the operation until it was all over, whereas heavy troop movements by the British Army immediately drew it to the attention of the media, with all the attendant political problems.

12 THE WORLD AT LARGE

The operations in Oman had lasted from November 1958 to January 1959. During that time my wife, whom I had left at our home in Malaya, had no idea where we had gone. Duke's death somehow got into the national press, so those we had left in Malaya knew we were in a shooting war, but only that it was somewhere in the Middle East – which covered a multitude of sins.

Rumours that four of us had been killed, including John Watts and myself, were quickly stamped on by HQ 22 SAS in KL, but not before Ann had suffered some aggro. After several weeks, Ann was told I would not be returning to Malaya so she was to pack all our household kit and prepare to return to Britain. My friend and his wife, who lived next door, helped with the packing. Having been put off flying by a rough ride on the last leg of her flight out to Singapore, Ann opted for a sea passage home and really enjoyed it!

'D' Squadron returned to the UK in March 1959, and enjoyed a couple of weeks' leave. Our new base was at Merebrook Camp, Malvern. It was a rather tatty old wartime camp, of the usual pattern: a few brick-built administration buildings, a lot of wooden huts.

The first day after returning from leave we did a twelve-mile route march with fifty pounds of kit in our rucksacks. We marched on roads. Practically the whole squadron was crippled with blisters and other problems. It struck us all as rather stupid that we, who could cope with some of the roughest terrain in the world with no one (that I know of) falling sick or raising a blister, could be destroyed by a bit of a walk on roads. We didn't even walk up the Malvern Hills; the route march was all on fairly level roads, east of Malvern. Later, when we tackled the Welsh mountains with heavier packs over much greater distances we came into our own again. I don't think any of us will ever forget the disaster of marching on roads.

At Malvern we were inundated with a plague of clowns. It wasn't so much their numbers as the positions of power they held. They caused a lot of problems, some of which dogged the SAS for years afterwards. The most memorable of these problems was known as 'bubbling RVs'. No, not a rendezvous in a bubble-bath. If you were 'bubbled', you were betrayed to the enemy. If, on exercise, your RV was 'bubbled' the 'enemy' would be in ambush there. Then you usually spent the rest of the exercise being interrogated and having a hard time in general.

As we were mostly very experienced under field conditions we were not easily captured on exercises, so some clown – or clowns – decided it would be a good idea for us to be captured, partly to give us the experience of quite nasty interrogation, partly to give the interrogation teams more work.

The troggery learns those lessons quickly and takes a long time to forget them! For years after that period no SAS patrol would give its correct position on exercises. Elaborate evasion tactics were used in relation to all RVs. Troops and patrols made a habit of fixing RVs at least a thousand yards from any which were sent by radio. An RV which was known to HQ was always avoided like the plague. It got so that even our own people couldn't find us.

One clown came to North Africa with the squadron on our first desert-training session. He couldn't really navigate but was good at pretending. We spent all one morning driving in a straight line out into the desert, everyone eating the dust of the truck in front. Then we spent all afternoon driving in a straight line back to base, again eating dust. I think he was afraid of getting lost.

Another clown who came with us at the same time insisted we should have our kit laid out on our beds when we were in base for a day or two. He actually charged a friend of mine for having his PT shoes the wrong way around on his bed. I heard a good story about that one after he went back to his unit. According to the tale, he was in charge of some missiles somewhere in West Germany, and when an inspecting team arrived unexpectedly one day, they found he had all his missiles aimed on a back bearing (the opposite direction to the original setting). They were pointing at Paris or somewhere like that. For which the clown got the boot from the army. Probably wishful thinking, but it makes a happy ending for a lot of people.

As far as I remember, it was at Malvern that 'the crab' first appeared. It was, I believe, 'A' Squadron's answer to our 'one-legged chicken'. The one-legged chicken was what stamped on all military equipment, leaving the well-known broad arrow everywhere it went. It only lasted a few months before being superseded by the crab, whose source I know not.

The one-legged chicken had many notable achievements, one of which was when its tracks were found leading into the Operations Centre, into the Ops Officers' Office and into that holy of holies – the Top Secret Safe. Where it laid an egg (real, large, brown and broken) with the word 'Oops!' chalked alongside. It had achieved that escapade in the dead of night – much to the horror of those responsible for the security of the Ops Centre and all its contents.

There were many types of crab at first. They all consisted of a simple drawing, an oval shape with six or eight legs, with a word or name in the middle. The difference being only in what was written. The challenge of the crab was to appear, large or small, in the most unusual or embarrassing places. It didn't leave tracks – just appeared. Like the one-legged chicken, the crab often pointed out cracks in the security system by appearing where there was no way it could. A few crab examples are worth a mention.

The regiment went to Hildesheim in Germany for training. Being too crippled at the time to parachute at night, I was sent on the advance party, by train and sea. We had a lot of time to kill between trains in London, and were in a mixture of recumbent postures on a platform at Liverpool Street Station.

Someone got bored and crabs started to appear. The Military Police on duty at the station took exception to: 'The next train from this platform will be the Sex Crab to Reading.' So two burly MP sergeants came stamping over to us to find the culprit. Needless to say, no culprits were found, but those two nearly had a fit when they noticed their boots after they walked away. On each of their highly polished toe-caps was a miniature crab.

At Hildesheim we were based in a barracks occupied by an artillery unit. In the barracks there was a fairly tall aerial mast, about one hundred feet high, I think. As daylight broke one morning, a huge cardboard box was to be seen at the top of the aerial. On each of its four sides was painted a huge crab.

Years later, talking to the gent who had been our RSM at Hildesheim, he told me that when he first saw it he grabbed the first SAS soldier he saw and sent him up there to get the box down. The man did so, without the slightest hesitation. It wasn't until after he had disappeared into the mess hall that the RSM realised it was probably the very man who had put it up there. There weren't many who would climb a thing like that.

One of the most famous crabs was on the end of a runway at Bahrain airport. It was so big, pilots claimed they could see it from eighty miles away.

Higher authority, often the target of the crab, eventually managed to fade it out – or perhaps it got 'demobbed'.

The regiment spent about a year at Malvern. It was a year of catching up on all those skills which are so much taken for granted now, but which up to that time had not been essential to the operational requirement, so had mostly not been touched. There were suddenly courses galore. No troop was ever at full strength, there were always several members away on courses of one kind or another. Although most of us could drive, very few had British driving licences so a massive driver training programme was instigated to put that situation to rights.

Long, flogging exercises in Wales seemed to fill in all the gaps. We hoofed from one end of Wales to the other, and back again, carrying vast loads as usual, but now aggravated by extra radios, a lot more machine-guns (Brens and .30-calibre Brownings) with piles of ammunition.

Parachuting became an ever-present threat to the smooth running of everything else. Being a very bad parachutist, I could never see the sense in jumping just for the sake of it. My landings were always hard and fast. I had what I called my dishcloth technique. I landed like a dishcloth – splat! There was no chance of doing the much-vaunted parachute roll. I did all the things I had been trained to do. It never seemed to make any difference. I rolled all right but there was nothing controlled about it.

There were not a lot of parachuting injuries in the regiment: nevertheless we always viewed jumping as a waste of time and potential hazard to future training and operations – unless it was linked to other training such as exercises, tests to improve DZ techniques or RV procedures. In fact, most of our parachuting was used to the full, to test anything and everything which could be linked to it.

My first injury came in 1959 at Weston-on-the-Green, near Oxford. It was another jump for nothing, of course. As I descended faster than most, there was always the danger of pulling one of the other people in on top of my chute. It had happened before, but at a height where things could be sorted out. This time I knew nothing about it. People who saw it happen say my parachute was completely collapsed by someone crashing down on it about fifty feet from the ground. All I remember was thinking, 'Here comes another fast landing.'

I was knocked unconscious by the landing and when I came to, the man who had collapsed my canopy was bent over me apologising. That was the first I knew of anything wrong with the landing. Nothing was broken but I had badly torn a lot of muscle in my stomach, groin and thighs. I walked off the DZ with a burning sensation where the muscles were damaged, and didn't get the full effect until it had all gone stiff during the truck ride back to Malvern.

The MO gave me the next day off, then I did my best to join in with training as I didn't want to miss out on anything. That injury played me up for months, whereas if I had rested it in the first place it's likely it would have mended in a week or two. On the other hand, if we hadn't done that stupid jump . . .

My second parachuting injury came after the regiment moved to

Hereford in 1960. We had an aircraft available for a day or two, using an old airfield at Madley, near Hereford, for the aircraft and jumping on to the Lugg Meadows, east of Hereford. On my second jump I had an extra-fast landing, ploughing up the turf as usual, but also busting two rivets out of my steel helmet. This time it was only concussion and, although I never knew it at the time, a hairline fracture to the base of the skull. I was over the concussion after a couple of days but suffered loss of memory on occasions for a long time afterwards.

One of those memory lapses was worrying. I had gone into town to buy something, driving my car as usual. I returned home, as I thought, only realising something was wrong when I found the drive gates closed, different curtains at the windows – and realised I didn't live there any more. I drove away, parked and sat there trying to remember where we lived. We had moved several months – maybe a year – previously. I knew who I was and everything else, except where my home was.

After half an hour of searching my pockets, the car and my memory, I hit on a plan. I decided I would get my address from the duty clerk at the camp. Luckily for me it all came back as I crossed the old Wye Bridge. Only my wife ever knew about that one.

I was knocked unconscious on several parachute jumps before I was saved by the introduction of the PX parachute in 1962 or 1963. I think the old X-type parachute just about kept me upright, whereas the PX actually slowed me down as well.

The regiment moved from Malvern to Hereford in 1960. Before we made the move Ann had become so pregnant we had to exchange the bike for a car. It always amazed me that she, who never batted an eyelid at speeds in excess of one hundred mph on a motorcycle, got into a right panic at anything over thirty mph in a car!

Soon after the move to Hereford, 'D' Squadron went to Kenya for two or three months. We were based at Nanyuki, which, although on the equator, is elevated enough to be quite pleasant, temperature-wise.

Kenya had been the scene of some particularly nasty terrorist activities during the 1950s, when the Mau Mau uprising took place. By the time we went there, however, the problem had been solved more or less to everyone's satisfaction and peace had been declared. Nevertheless, a few of the hardcore terrorists, mostly those wanted for murder and other atrocities committed during the Mau Mau troubles, were still known to be semi-active and hiding in the bush, notably in the denser bush and jungle on the Aberdare mountains. There was very little chance of our making any contact with the terrorists on our training but it did give an extra kick to things when we were working in the Aberdares.

One of the first things I remember at Nanyuki was the flu epidemic. Asian flu had struck in the UK. Two of the squadron had it when we flew out to Kenya. A week after being shut in an aircraft with them we all had it! All training came to a grinding halt.

During that week or two, while the squadron was more or less laid up, I went on a snake hunt around the camp perimeter. The MO had said he would be interested in any snakes we might find. I found a snake and, with help from a friend, caught it and took it to the MO. He was not too grateful, as it turned out to be a rather large puff adder, reputed to be very dangerous in the venom department.

In Kenya, 'D' Squadron was very fortunate to have one of the best squadron commanders, Major John Slim, whose forward thinking and strong faith in the men he commanded was to make a long-lasting impression on the whole regiment.

We threw away the book on jungle experience gained in Malaya and started with a clean sheet. We tried new ideas in the African jungle, some of which were old ideas not usable in the Malayan jungle. Nothing was taken for granted. The squadron didn't waste any time in Kenya. Some things worked, some didn't, but mostly we finished up knowing which would and which wouldn't.

We tried marching through bush on the slopes of Mount Kenya at night. Part of our route was through fairly open bush, part was through quite dense jungle. A lot of heavy, ground-shaking animals were disturbed that night, but I have no idea what they were. One small incident, however, put me off night movement in thick jungle. We were moving fairly quickly so were not very quiet at the time. I was leading scout. Suddenly the ground gave way under my feet and I fell forward. Luckily for me the sling on the machine-gun I was carrying caught on something and I hung on to the gun. When the other people reached me I was dangling in space over a good forty-foot drop to jagged rocks. It was a ravine about twelve feet wide with a tiny stream at the bottom.

Every morning when we were in camp we had PT (physical training) before breakfast. Half was normal army PT, the second half was always unarmed combat training. Morning PT became interesting.

In a bomb crater on the slopes of Mount Kenya, we built a 'house' with sandbags and this, as far as I know, was the first ever CQB (Close Quarter Battle) House for 22 SAS. Although the Grant Tyler methods (aiming with the body and shooting from the waist) with which we were trained in Kenya have long since been superseded by continually improving, better methods, it was a good start to CQB shooting and was the start of the regiment's ability to successfully go into places like the famous embassy in London.

The Aberdare forest in Kenya was one of the worst jungle areas I have ever seen. The trees are often not dense enough to keep out the hot sun. Attap, the curse of Malayan jungles, was not present but was more than made up for by – would anyone believe? – blackberry bushes and monster stinging nettles. We seemed to wade through oceans of tearing, stinging mixture of them.

There is also a lot of uphill in the Aberdares. 16 Troop never claimed to be the world's lightest sleepers, but very little moved near us at night without all or most of us knowing about it. We were all a bit shamefaced one morning in the Aberdares though. Two or three elephants had walked through our deep jungle camp in the night – and no one had heard a thing. The great footprints were there for all to see in the morning. Someone claimed those bloody nettles had drugged us.

Most of us came face to face with big wildlife at some time or another in Kenya. None of us seemed to get the urge to shoot them though. I was leading scout for the troop most of the time. All leading scouts carried Bren guns, because of the possibilities of upsetting the wildlife. Our reasonably quiet movement caused us to come upon the animals rather unexpectedly but, as far as I know, there was never any cause to use weapons. If the animal didn't back off after a few seconds of 'eyeball to eyeball', we walked quietly around them. There seemed to be a sort of mutual respect. It was cowardly on my part really, as it's easy to walk round huge unarmed animals when you have a relatively powerful machine-gun in your hands.

The more I saw of the wildlife in Africa the less respect I had for the big game hunters. The animals are so easy to shoot, it's pathetic that anyone could take pleasure from it. A mother elephant running forward to put herself between her calf and possible danger can be a bit unnerving, but we had to respect the normal maternal instincts – just standing still, making no threat, is usually enough to satisfy the great animal there is no danger.

There is only one big game worth shooting, and that's the one who shoots back. For voicing those sentiments and stating it would be more difficult – and raise more interesting reactions – to shoot cows in a field back home, two of us were politely requested to leave the bar of the Equator Hotel, near Nanyuki. The bar was about half full of big game wounders at the time. I think the management were worried about damages.

A couple of weeks before the squadron was due to return to the UK from Kenya, a message came saying Ann was having problems having our first child, so I was flown home right away. In the event, there was no problem with the birth, but the doc said it was because I had arrived home. For all the help I was, I think he was joking.

So now I had a son and a lance-corporal stripe (which had somehow arrived in Kenya). Things were getting more complicated.

Although we were based in Hereford, we hardly saw the place. 'B' Squadron had been disbanded after the regiment reached Malvern, so now there were only two operational squadrons. The two Sabre Squadrons rarely saw each other. It was very seldom we were both in base at the same time. In fact, as I remember it, it was very seldom any of us were in the Hereford Base.

During one of our many exercises in Wales there was a small happening which I cannot leave to be lost in the mists of time. We were using a military establishment at Towyn, North Wales, as an exercise base and interrogation centre. The place was also the base of a boys' regiment (artillery I believe) and on the day of one of their passing out parades, when all the proud mums and dads had come to watch their smart lads on parade, one of our lot made a bid for freedom from the interrogation centre.

Imagine the distraction for the proud parents as this dishevelled, unshaven apparition, smartly turned out in his underpants (known in the army as 'drawers dracula'), sprinted across the parade ground, hotly pursued by several irate gentlemen dressed in Eastern Bloc uniforms (red stars and all!) waving Russian submachine-guns. I would have hated the job of explaining that lot.

In that period between returning to Britain in 1959 and about mid-1962, we spent all our time living out of rucksacks or suitcases. As an example of this continual bouncing from one job or place to another, I have always remembered the first seven or eight months of 1962. In those few months I saw thirteen foreign countries, and later that year I went to Scotland for the first time in my life.

Some of the training we did was a dreary flog but most was interesting, often challenging. One such challenging session was our first exercise in Denmark. In my case it was more so, as it was the first big exercise I had as a patrol commander, and with very little time to prepare as I only arrived back in Hereford the day before we left for Denmark.

Four of us had been sent to the Snowdonia area in north Wales to advise and help train the Staffordshire Regiment in mountain warfare. We arrived back in Hereford expecting a few days with our families. I was sent for by the squadron commander and told I would be taking a patrol to Denmark the next day.

There followed an intensive briefing with piles of maps, air photos and sketches. The next day we flew to Sylt, an island off the Danish coast. The following night we parachuted on to the mainland. It was one of the few times when my blind faith in the RAF was shaken. There were five of us in the patrol, a Marine Commando had been sent with us for some obscure reason. ('Obscure' because he was a frogman, and my target was nowhere near any water!) Perhaps the navigator mistook metres for feet, I don't know, but we were dropped from three hundred feet. Only two of the patrol had released their equipment before we hit the ground. Luckily for us, the Drop Zone was covered in thick springy heather so no one came to much harm. A few bruises and a bit of a shaking was all. Being a supreme coward I had released my kit just in time. When you are jumping into a black void and can't see a damn thing there's no sense in taking chances.

The main purpose of the exercise was to test the Danish defences, who certainly made life difficult. But, with a liberal helping of good luck, we managed to hit our target right on the button and get clean away. Not all the patrols were as lucky and some spent a few days not enjoying some very nasty interrogation. The interrogators were Brits, especially sent over for the job. They were – to say the least – most unkind. The training was for them as well as us, and they made the most of it.

SAS night jumping in those days was always a bit hairy, to put it mildly. Not so bad if we were jumping into somewhere in the UK. But if it was somewhere on the continent there would usually be a long uncomfortable flight with stacks of time to dwell on all the wrong thoughts. There was always too much noise in the aircraft for much conversation, so the best bet was to sleep, as the lighting was not good enough for reading.

Usually there would be about thirty of us in the aircraft, sitting on hard metal or canvas seats down each side, facing inwards. So a certain amount of entertainment was possible by watching the more fortunate sleepers with their lolling heads, open mouths and twitching. On a long flight our parachutes and equipment would be stacked down the middle between us. The aircrew would let us know when it was time to start getting organised.

It was most likely we would be dropped in patrols of three or four, sometimes more, on several different drop zones (DZs). These were often what those at the top of our hierarchy called 'sporting' – somewhere where no one in their right mind would expect paratroops to land. The most sporting one I can remember was a meadow alongside a large river, about three hundred yards long with overhead powerlines just outside one end and a large timber yard at the other. We were issued with lifejackets for that one. The comments were rife about what to do with the lifejackets if we found ourselves in the powerlines or timber yard! Thanks to the expertise of the RAF there were no problems.

DZs for normal airborne operations which involved troops being dropped by the plane load, were normally at least eight hundred to a thousand yards long and five hundred metres wide.

Some of our patrols had DZs which were much more sporting than any I saw but – what with the vagaries of the wind and the dependence on absolutely perfect navigation and timing by the aircrew – they were very welcome to them.

About twenty minutes before reaching the first DZ we strapped on our parachutes, then sat down again to clip on our heavy equipment. Soon after that the first patrol got the order to stand up. The RAF dispatchers took off the door, leaving a gaping black hole into the night through which came the deafening roar of the engines and a howling cold wind to half-freeze everyone in the compartment. The main cabin lights were switched off, leaving us all in very dim, ghostly indirect lighting which was just enough illumination for the patrol to do its final checks. The lack of bright lights helped our eyes get accustomed to the dark when we left the aircraft and had to cope with the blackness of the night.

It also meant the aircraft was blacked out and difficult to see as it descended to roughly six hundred to eight hundred feet ready for the drop. As well as descending, the aircraft slowed to about one hundred and twenty mph, all of which – along with a few last-minute flight corrections – often produced a bit of turbulence. The patrol, standing in line behind the first man, who was about one yard from the doorway, were often swaying about like drunks, with nothing to hang on to and weighed down by anything from one hundred to one hundred and eighty pounds of parachutes and equipment (often more).

When the red light came on over the doorway the dispatcher screamed: 'Stand in the door.' The first man stepped into the doorway ready to jump and the patrol closed up behind him. Within half a minute the green light replaced the red and the dispatcher screamed 'Go!' and slapped the first man on the shoulder, and one after another the patrol put their lives on the line and threw themselves into the black, roaring slipstream and the howling void of the night – hoping to Jesus Christ everything worked and the Air Force were right on the button with their navigation! (No satellite navigation systems in those days.)

For the first few seconds outside the aircraft you were a helpless rag doll. You tried to keep a good jump position but were hammered by the slipstream and wind while dropping over one hundred feet before the chute opened. You felt your parachute had opened, so you looked up to see if it all looked okay. Sometimes you could see the canopy against the sky, sometimes you couldn't. You also tried to see if the lift webs and lines looked right and were not twisted.

Being a supreme coward, I could never get rid of my equipment quick enough. Release the leg strap, quick glance below (in case of friends), then flip the hooks – and let her go! I hated the thought of hitting the deck with that lot still hanging on me. My landings were savage enough as it was.

The noise of the aircraft rapidly faded into the night and a comparative silence engulfed the immediate area. You were often hanging from an unseen parachute and rushing towards an unseen earth. Unreal! But it didn't last. You hoped you were in the right place. You must not steer or the patrol could be scattered and delay organisation on the ground. That's a laugh, you could rarely see where to steer, and those old X-Type chutes would hardly steer anyway.

So you hung there and waited for it, trying desperately to see something – anything. Kept your best landing position and listened for your kit hitting the ground about fifteen feet below you – that gave you about half a second of warning.

Then crash! and you were on the ground, rolling on to your stomach to grab the parachute lines and collapse it before it pulled you into trouble or was seen by the opposition. Pull it in and roll it up. Bless those girls at Upper Heyford who packed it and gave you a safe ride down.

The patrol would normally be on its way off the DZ within ninety seconds of landing.

While on the subject of parachuting, one of Big Doc's exploits is worthy of mention. Doc is about the same height as me and so had the same problems with low-exit doorways. The happening came about because Doc's solution was different to mine. When in the number-one position – first out – Doc would stand with his head outside the aircraft when ready to go. He was obviously oblivious to all except the engine and slipstream noise.

We were flying from somewhere in Malaya to jump somewhere on the northern part of Singapore Island. Doc was standing in the door as we approached the Johore Straits: a very narrow strip of sea between Johore and Singapore.

The dispatcher must have tapped his shoulder while trying to see out of the door – and Doc was gone! With the luck of the Irish, Doc landed in a nice soft paddy field in Johore and not in the shark-infested waters of the Straits, while the rest of us flew on to smash down on to a hard, sunbaked DZ on Singapore Island.

Once, when we were jumping on to a vast snowcovered DZ in cloudy moonlight, I found myself looking down in shock at a huge, thick wooden fence right in my way. Struggling to steer away from it, all I could think was, 'Where the hell did that come from in a place like this?' Having climbed halfway to the canopy and achieved practically nothing

I realised the 'fence' was only two parallel, overlapping vehicle tracks.

One of the most memorable training stints was in Oman. The one thing which stood out above all others to make that trip unforgettable was Charlie, a gifted trooper in 16 Troop. We were based at Ibri in northern Oman, and the accent was on desert training with vehicles. The vehicles we used on that trip were mostly Austin K9 one-ton trucks, plus a few Morris Commercials, which were also one-tonners.

We perfected our desert navigation techniques and learnt how to get vehicles over any desert terrain. The training was hard work but very enjoyable, as can be imagined. It was all very hit and miss at first, but after hours spent digging out trucks we soon learnt how to recognise soft sand and how to cope with it.

Sharing our camp at Ibri was an old sergeant, or staff sergeant, in the Royal Engineers. His job was to drive along all the roads in his area every day, in an especially equipped, but very battered, Land Rover, checking for mines and lifting any he found.

How he got away with it I've no idea. The detection part of the equipment was held out in front of the Land Rover, about twelve feet from the radiator on tubular steel poles, when it passed over a mine it flashed a warning light in the vehicle. But the Engineer bloke drove at about thirty to forty mph. There was no way he could stop in twelve feet. A lot of our people claimed he was only twenty-one – it was his job which gave him white hair and made him look forty-odd! His two or three rows of Second World War campaign ribbons must have meant something, though.

I don't know where he was the morning one of our trucks drove out of the camp and hit a mine, just twenty yards from the gate. One of our people, a passenger in the cab, thought they had a tyre blow (so he claimed) until he saw the three-foot deep crater in the road and bits of wheel scattered around the area. Being only about forty yards away, cleaning my teeth, I got showered in gravel and nearly swallowed my toothbrush. A few trucks were blown up on mines laid by the local rebels, but no one was seriously injured, so all it did was add a bit of spice to the trip.

Charlie has been called a genius. He was – at the very least – a superb mechanic. His mechanical exploits on that trip to Oman were legendary in the SAS. I can't pretend I knew what he did, or how Charlie solved our problems, but we all saw him perform – what were to us – miracles. He had less desert experience than most of us, so had not had to solve these problems before.

One of the first things he showed us was that because water cannot be compressed, a one-ton truck could be jacked up on a full jerrycan of water. This helped us a lot as solid, strong load spreaders to put under a jack in soft sand are not easily found in the areas where they are most required. We tried it with full jerrycans of petrol, or at least Charlie did, while we stood at a very safe distance. It worked, but, because petrol can be compressed, the jerrycan looked a bit sick afterwards. Not recommended! The jerrycan has to be opened with extreme care if the petrol is required later.

In the intense heat of the desert the trucks were continually stopping owing to lack of petrol reaching the carburettor. At least, that's what Charlie decided. After a day or two wrestling with the problem he decided drastic action was the only answer. We were a couple of hundred miles from base and Charlie had only his toolbox to work from. Somehow he drilled the filler cap on one of the Austins and fitted an air valve, then rigged an air line from the spare wheel so that air pressure went to the fuel tank and pushed the fuel through faster to the engine. He did other things, I suppose, to make it work, and it did. Charlie's modified truck had no more problems.

The only problem was that we started running out of spare wheels as the more distance we put in, the more punctures we had, so Charlie decided the system needed a rethink. His eventual solution was to pressurise the fuel tanks by pumping air directly into them. We watched from the safety of a distant sand dune as Charlie went to work. We saw him drive the truck madly around, this way and that, for a while to test it. Then he came over to us. His pressurised fuel tank had changed shape. It was now very rounded, bulging as if about to explode. Charlie assured us it was all safe enough, and proceeded to modify all the troop vehicles. It worked really well and only needed more air twice every day during constant travel. That was easy compared to breaking down every ten to fifteen minutes.

Charlie's real stroke of genius came at the end of that trip to Oman. His endless search for better ways to do this or that made life a little uncertain at times and occasionally landed us all in trouble, but he was well loaded on the credit side.

We were to be flown out to Bahrain aboard a RAF Beverly from a patch of desert near Ibri. The Beverly was an awkward-looking four-engined aircraft which always appeared (to me at least) to defy all the laws of gravity and aerodynamics, but it could land or take off almost anywhere. Our Beverly landed fine, then taxied to where we were waiting to climb aboard. As was normal practice, three engines where shut down and one left running, so as to start the others when required. The engine which had been left running suddenly gave a couple of splutters, then stopped.

There was a sudden great silence. We realised, without being told, there was no way that Beverly could start itself. The aircrew said they would radio for a starter to be flown from Bahrain, but it would be unlikely to reach us before the next day. Breakdowns and delays never happened coming out, but we were quite used to problems when returning to the UK, so we resigned ourselves to another night in the desert.

Charlie had a chat to the RAF flight engineer, then rushed off in a Land Rover. Word spread that Charlie had an idea. I had a quick vision of the aircrew standing around the burning wreck – but quickly dismissed it as pessimistic thinking. Charlie returned with a lot of telephone wire plus a few odds and ends. He and the flight engineer climbed into the aircraft.

After a few minutes Charlie emerged and got to work on two Land Rovers, wire going in all directions. Then, with the ends of a lot of wire in both hands, he climbed back into the aircraft. A few minutes later he appeared in the doorway and signalled the drivers to start the Land Rovers. They started up and revved the engines hard. When he was satisfied with the noise Charlie turned and gave the thumbs up to the flight engineer, somewhere inside the aircraft. The next thing, an engine turned over, fired and started. The engine noise was nearly drowned by the cheering.

Part of our training for desert vehicle-borne operations was taking part in a squadron exercise, which involved all patrols making an RV somewhere a long way out in the desert, then moving as a squadron to put in a night attack on the Fahud Bowl. The Fahud Bowl is a large circle of rocky hills, the remains of a volcano, sand on the inside, sand on the outside and several miles across.

The navigators among us, notably a couple of the troop sergeants were, I considered, smack on target as we started our approach from perhaps forty or fifty miles out in the desert. We maintained a good speed over mostly gravel plain. All was going according to plan when the clown running the show (not the squadron commander, who was elsewhere) suddenly decided he could see the black jagged outline of Fahud away to the left of our approach bearing.

The squadron came to a halt. There was a quick 'Chinese Parliament' with all the good navigators insisting on maintaining our previous direction and the clown insisting we turn towards the black outline on the horizon. The clown, of course, won.

The incident was long remembered as 'the time so-and-so tried to put in a squadron attack on a cloud bank'. Nevertheless, that particular clown impressed me a great deal that night – by being able to laugh at his own mistake and call himself a 'bloody fool' in front of the squadron. A clown who is a bloody fool and not afraid to admit it to his fellow men is a great improvement on the usual lot.

That particular clown has since become a damn good soldier, impressing a lot of people with his sheer guts and determination. If there are any qualified to write a book on what makes things tick in the SAS, then he must surely be one of the few.

During that trip to Oman there was some problem over the Baraimi Oasis, which is on the border of northern Oman and had been a disputed territory for many years. Saudi Arabia had some idea of taking over the Omani territory and possibly the Trucial States territory, thereby claiming the whole of Baraimi Oasis which was shared by all three countries. I never knew all the ins and outs of the situation, but someone called us in to help out.

A large contingent of the Saudi Army had arrived at Baraimi to back up their claim, so some of us, I think about two troops, were asked to make like a large military force in the desert about two miles east of Baraimi.

We had about eight trucks and less than thirty men but we went to work with a will, drove like hell all over the place making huge clouds of dust when we arrived, then kept up the dust cloud until dark. We then got all the tommy cookers we could spare and made dozens of little camp fires all along a low ridge so they could be seen by the opposition.

Next day the Saudi unit had gone, so we went off and got on with our training. Whether or not we had any effect on the situation we never heard.

Some of our people went off to get experience of desert patrolling on camels, something I had no ambition to do. I like animals, but for some reason I could never get on well with camels. Luckily I didn't get 'volunteered' for the camel patrols and was sent with two trucks to check on a salt mine in the desert, about two hundred miles south of our base. Having no idea what a salt mine looked like, we used our imagination and, of course, got it all wrong.

Arriving at the correct latitude and longitude we could see nothing in the way of mining activity, so did a lot of searching around in a vast expanse of hot, flat desert. There were two or three small rocky hills a few miles apart, so two of us climbed the largest one (about three hundred feet) with the idea of getting a good look around with our binoculars for the elusive salt mine.

Reaching the top of the hill we found it was hollow, like a volcano, and that was the salt mine. It was at the bottom of the 'volcano'. The miners were three old men and a camel. They were mining rock salt, and not very much at a time. Every few days they sent out a load of salt on the camel, which brought back their supplies of food and water on its return. They were a long way from any habitation and I couldn't help thinking it would be rough on them if the camel was injured on one of its trips.

There was a crack in the side of the hill, through which the camel could just squeeze when loaded. The desert wind hid the occasional tracks made in the sand, so making it one of the best-hidden salt mines in the world. No wonder we had problems finding it.

On one of our trips to the Persian Gulf area someone bogged up the aircraft booking and we didn't have enough seats in the aircraft to fly home the whole squadron. It was just before Christmas and there were hopes of getting Christmas at home.

The squadron commander decided we would draw lots for which troop stayed at Bahrain. So the four troop commanders (I must have been a sergeant by then) drew straws and, as usual, I lost. The troop was disgusted, but that didn't help, so off went the squadron – with a lot of cracks about saving us some Easter eggs – leaving us to find our way home any way we could.

I went to the RAF types who know all the answers about what flies to where. They got us fixed up with a lift that night, on the milk run Comet. The only snag was it was via Aden, Khartoum and Malta – the long way home – and would take two days. When we arrived in Aden, the aircrew realised we couldn't land at Khartoum as it was one of those places that objected to foreign troops in uniform on its soil, so we either got civvy clothes or took our chances on another aircraft from Aden. We had about half an hour to solve the problem.

We were all stood near the passenger terminal building, stony-broke and taxing our tiny minds for a solution when it came walking right up to us. An aircraft from Britain had just landed, bringing dozens of white-faced young lads for their first tour of overseas duty. And they were all in civvy clothes! I don't know what happened but suddenly we were back on the Comet in rather tattered civvies.

The best part of that story was we arrived home days before the rest of the squadron. Their aircraft developed a problem and they spent four days in El Adem, or somewhere beautiful.

The regiment was into the business of exploring all possible types of military operations and free-fall parachuting was beginning to catch on. My friend Keith, the shotgun artist, was killed on Salisbury Plain when the regiment took the British high-altitude free-fall record. He didn't pull his rip-cord. No one seems to know why. In memory of Keith, a clock was purchased to stand in our UK base. The clock tower is inscribed with the names of men killed while with the regiment on training and operations. When things looked a bit dodgy sometimes, we used to joke about beating the clock. Considering the training and operations over the years, there are not that many names on the clock. Still too many, though. Free-fall parachuting was something I had no ambitions with. Mother Earth was still too fond of me and rushed me back to her too quickly.

I had a slight brush with the watery stuff when I was sent to Devizes to have a go at the Devizes to Westminster canoe race. I was a dismal failure. The others had built up to paddling the twenty-five miles from Devizes to Newbury before I arrived, so I had to paddle that twenty-five miles the first time I ever stepped into a canoe. We did it three days on the trot, then started on the stretch of canal from Newbury to Reading. I did the stretch twice, I think, before the old injuries put a stop to it. I had thought canoeing might be a good thing to strengthen my weak arm and shoulder, but I think I went too far too soon. I stayed with the canoe team, driving the safety vehicle, until they finished the race, beating all previous records. One of my friends, who took the record for his class, lost a stone in weight in the twenty-four hours it took to win.

One of the worst experiences I can remember involving water was something we did in Hereford – to impress a visiting general. It was midwinter, we had just returned to the UK from somewhere hot (climate-wise). It was decided a demonstration river crossing would be a good thing for the General to watch. On getting lumbered to be one of the four-man patrol, I could think of a lot of other things the General would be interested in, but there it was – we were stuck with it.

A practice session the day before the General's visit went quite well, in spite of the freezing conditions, but overnight the weather changed for the worse. What might be called a cold snap set in. We did the crossing somewhere upstream of the old Wye Bridge at Hereford, with all our operational kit and weapons of course. It was an early-morning job. There was a bit of mist and when I got my first sight of the river I nearly died of shock. It was well up, as it usually is in winter, there was hoar frost everywhere and ice on the river, going out about ten feet from both banks.

Alex was first to go. I really felt for him as he smashed the ice and dropped into the water. Then it was my turn. Believe me, it was bloody cold. I never thought I'd reach the far bank. My whole body seemed to seize up with the intense cold. Brass monkeys would never have made it intact! It was touch and go whether we would! But everyone crossed safely and moved off from the far bank as an organised patrol, under the eyes of the brass.

Once out of sight, however, we were given a goodly shot of hot toddy, wrapped in blankets and rushed back to camp for a quick change. A hot shower never felt so good.

Sometime in 1961, 22 SAS decided to have each troop in a squadron specialise in a particular type of warfare. Personnel were shuffled according to their abilities. 16 Troop, which I had been with since joining SAS, became 'D' Squadron's Free-Fall Troop. 17 Troop were made Boat Troop, 18 Troop Rover Troop and 19 Troop became the Mountain Troop. Almost every man in 16 Troop was transferred to 18 Troop. We went to work on all types of vehicle operations; desert navigation and all types of cross-country driving in Europe and North Africa. We all trained on mortars and machine-guns and mastered the art of bringing jet fighters and bombers to attack targets.

The Free-Fall Troop seemed to do very little except free-fall, at which they became pretty good. The Boat Troop worked with the Marines quite a bit getting to know the ropes with all things to do with water: beach reconnaissance, submarines and diving. Mountain Troop went climbing mad and climbed everything in sight. They also specialised in Arctic warfare, ski-ing and all those kind of activities.

Nevertheless, we all had to keep in touch with our jungle skills and rarely went long between Bergen humping of one kind or another. There was always a damn great mountain to climb somewhere.

Between 1959 and 1962 several instructors' courses came my way. I became an instructor in heavy drop, the parachuting of vehicles and equipment, and airportability, the loading and unloading of vehicles on different types of aircraft. I also qualified as an instructor in astro-navigation, first aid, nuclear, bacteriological and chemical warfare, air-photo reading and desert navigation. The only course I didn't enjoy was nuclear and black treacle. I did that one just after getting a bad concussion (parachuting) and couldn't concentrate or remember anything. I scraped through, but only just.

We all did a lot of other training and courses on a wide variety of subjects. Parachuting reared its ugly head on several occasions, but somehow I survived, never injured enough to stop me from sensible training or operations. The 'heavies' in the squadron were always interested in ways of slowing their descent by parachute. There was a theory that a really heavy load would contribute to a soft landing. When jumping with equipment, we dropped it to dangle on a line fourteen feet below us after leaving the aircraft. The theory was: if the equipment was really heavy it would cause a faster descent until it hit the ground. Then there would be a good braking effect on the parachute due to the extra air pressure, before our feet touched.

There was so much argument over the pros and cons of the theory, I decided to try it. We were jumping on to Hereford racecourse. My kit weighed something in excess of two hundred pounds. I could hardly lift the damn thing when it came to standing up ready for the jump. One of my friends, 'Big Doc', was behind me in the stick to help with the exit from the aircraft, as there was no way I could make a good exit with all my gear and I certainly didn't want extra complications.

The exit went well, with a massive push, well timed by Doc. Then I whistled down like a bomb, wishing all the way down that I wasn't so thick! So the theory was correct on the first part; a definitely fast descent. The rest of it wasn't so good. I wasn't hurt, but that bloody load towed me into the ground like a train. There must have been some brake effect or I'd have been injured, but it was the only time in my life I said, 'Jesus Christ!' before I hit the ground.

The great theorists, some of whom had landed previously and watched my effort with great interest, now argued it needed a longer line to give the brake effect more time to work. Others claimed the answer was to jettison the pack at about forty feet.

My answer was, 'Next time – you bloody try it!

On New Year's Day 1962, two of us from 'D' Squadron and two from 'A' Squadron flew out to Bermuda to train a local militia unit: the Bermuda Militia Artillery. They were no longer an artillery unit and were trained as infantry. There were two militia units in Bermuda. The BMA was all-black personnel and the Bermuda Rifles was all-white personnel. Why they were segregated we couldn't figure, but they were both happy enough with the arrangement. We found the BMA was a damn good crowd to work with. They had a typical British Army sense of humour and the training went well.

One small incident stands out in my memories of Bermuda. I was using ten of the men to give a demonstration of fire and movement to one of the companies. The Bren-gunner was about six foot and six inches, had arms like my legs, muscles on his eyelashes and made the light machine-gun look like a toy in his hands.

All went well at first, then the Rifle Group were firing and the Bren Group weren't moving. Before I could stop myself, I shouted, 'Come on, you black bastard – move your arse!' The monster came tearing up from his fire position, waving the machine-gun in one hand, a big grin on his face, shouting, 'Okay, you white bastard – I'm coming!'

After that it went like clockwork, but I felt about two inches high – and wished I was less. The thing is, I would have shouted the same words if he had been white, red or blue. There were times when the whole bloody army called anyone 'black bastards' in those days. I meant no malice and there was certainly no thought of racism on my part. Luckily the company was in good humour as usual and took it all as a joke.

We were each given a small moped with which to get around the sights and scenes of Bermuda. Being the only biker among us, it did my pride no good at all to be the only one to fall off. Apart from feeling a bloody fool there was little damage. A gashed knee and a few holes in a good suit.

During the last few days we were in Bermuda they had their annual shooting competition, a sort of miniature Bisley, when all their rifle clubs and armed forces teams entered for various cups and prizes. The competition was open to any forces based on or visiting Bermuda – just about everyone who could be there in fact. There were teams from the US Air Force, Navy, Marines and Rangers. The Royal Navy, the Royal Canadian Navy and possibly other visiting or Bermuda-based forces.

Our friends in the Bermuda Militia Artillery insisted we enter in the Falling Plates competition, saying, 'Okay, you four – pick a four-man team for the plates.'

Being put on the spot and at the last minute, we were not too sure of our abilities with their old .303 rifles – which we didn't have time to zero properly. But there was no way out so, in for a penny, in for a pound, we did our best.

Our erstwhile trainees did better than they had ever done before, reaching the semi-final, where they were eliminated by a team from the US Rangers. This left us to face the Rangers in the final, an unenviable prospect, so we decided to play it our way.

The competition is quite simple: twelve plates to be knocked down by each team. Each man has ten rounds to fire. The ammunition is placed two hundred yards from the targets, then the teams move back to five hundred yards from the targets and at the signal to start, rush forward to the ammunition, load their rifles and commence shooting. The winners being the team which downs all their plates with the least ammunition, in a time limit of two minutes from the start at five hundred yards.

Throughout our stay, we two from 'D' Squadron had taken some stick about our unflappable laid-back approach to all things military. The BMA troops had enjoyed the difference but found some things difficult to understand.

Several hundred people were watching this final with great interest, including nearly every man from the BMA. From the five-hundred-yard point we were looking through an avenue of people to where our ammunition lay. At the signal to go the Rangers and our 'A' Squadron types took off like hares. We two commenced a steady walk. The crowd went mad. In a couple of places along the way, much to my amusement, I heard the words, 'Come on, you black bastard – move your arse!' It wasn't just the words that made me laugh, it was also the way that one of them mimicked my West Country accent so perfectly; he was a born comedian.

We arrived at the firing point with our breathing unaffected. I think I knocked down two of the plates before my friend had demolished the rest. He is a damn good shot and the credit for our win is all his, although 'A' Squadron had done their share. But we were in the business of training people and it was a good demonstration of the merits of unflappability.

We were exceptionally well treated in Bermuda and made a number of good friends. After five or six weeks we flew back to UK, spent one night in Hereford, then flew out to Malaya where the squadron were doing some jungle-bashing. I spent six weeks in Malaya, two of which I was again instructing. This time four of us were sent into the jungle with a crowd of Military Police from Australia, New Zealand and Britain to introduce them to the art of jungle warfare. One of the Aussies managed to shoot himself in the leg; otherwise that was another uneventful session.

The Malay Police Field Force provided me with the best couple of days on the trip. Two of us had to deliver some radio spares and a few odds and ends to a troop deep in the jungle. We travelled most of the way by boat. Two boats, in fact, supplied and crewed by Police Field Force personnel. At first we were on a fairly big river and the speed was welcome. It looked like being a pleasant trip. The boats were made of wood, long and slim with a twin set of forty hp Johnson outboard motors giving ample power.

Then things began to warm up. The river became narrower, there were rapids and rocks, swirling water and spray everywhere. The police folk obviously knew their job and were familiar with every ripple on the river. They must have been, as the pace hardly slackened. We were travelling upstream and I, for one, found it harder and harder to appear relaxed and nonchalant in the face of what appeared to be imminent damp death. I sneaked a glance back at my friend in the other boat. A white face and sickly grin did nothing for morale.

At some stage, when I could chance letting go of my hold on the bouncing, twisting boat, I tied my rifle to my left wrist with a piece of para cord. By that time I was working out my route back to civilisation from wherever the inevitable happened – overland of course! But it never happened. We went under trees, over trees, scraped the sides and bottom of the boat on jagged rocks – which seemed to come at us in an endless procession of near destruction.

In a couple of places, at least, we had to go through shallows when avoiding obstacles in the deeper water. The crew aimed the boat at the selected spot, went flat out, then lifted the engines at the last minute so we slid through a few inches of water for several yards into deep water again. After we turned into a smaller river the speed slackened. In several places we had to jump out and push the boat through shallows.

The trip up the river took nearly a whole day. We two Brits were damn glad to get on our flat feet and hoof to the RV point when the boats could go no further. Two days later we rejoined the boats for the trip back to base.

Coming upstream had been scary. Downstream was bloody terrifying. I won't even guess the height of some of the waterfalls we flew over, but when you fly over the edge and know you are going to land in froth – because that's all you can see – it tempts breakfast to the tonsils.

I consoled myself with the fact we'd made it upstream okay – when I'd thought there were heavy odds against it – and tried to enjoy the 'flight'. The downstream run, back to base, was completed in three hours less than the upstream run. Nuff said!

In the last few weeks of the squadron's training in Malaya, three things happened which were unusual. It must have been 'Animal Madness Month'. In one instance a patrol was charged several times by a rhino, only succeeding in breaking contact when they fired several rounds over its head.

Within a few days of that, an elephant stomped a troop base camp into the deck and tree'd a certain friend of mine for several hours. I wasn't there, but witnesses swore there was no way anyone could have climbed that tree under normal circumstances. It must have been a big tree or the jumbo would have shaken him out of it. Trees don't usually have branches lower than about forty feet in the jungle.

The third animal happening didn't concern our people; it happened to a Gurkha sentry. He was attacked by an elephant, so shot at it. According to the report we received, the elephant took the sentry's rifle, smashed it against a tree, then threw it away, walking over the Gurkha as it did so. The Gurkha had only a broken finger, which had been through the trigger guard of the rifle when it was snatched. He was otherwise unhurt.

It seemed strange that there were these contacts with large animals all in the space of a couple of weeks, when the regiment had operated in Malaya for years without such happenings. There must have been a few brushes with the big ones during those years but I only ever heard of one: a corporal in 'D' Squadron bent his shotgun barrel over a bear's head when they met unexpectedly over a fallen tree. The bear ran away holding its head with both paws.

Before leaving Malaya we spent an afternoon in KL, so I decided to go and have my favourite nosh at my preferred eating house, the Rex on Batu Road. I had a drink at the bar then sat down at a table. Recognising the same head waiter when he approached my table, I just said, 'The usual please.' He didn't bat an eyelid. Five minutes later I had my usual in front of me. I hadn't been in the Rex for over three years but they hadn't missed a thing, it was all there – T-bone steak, double roast pork, double mushrooms, tomatoes, two fried eggs, four sausages and chips. Those people are true professionals at their job, without a doubt.

After the training in Malaya we returned to the UK. I went on several more jobs until somewhere in mid-1962 I was posted to 23 SAS, a Territorial army unit with headquarters in Birmingham. My time with 23 SAS was rather mundane after the mad rush of 22 SAS. I found the part-time soldiers trained hard when their clown department didn't get in the way. Most of their training was organised by Paddy Nugent, who did an excellent job of keeping up SAS standards. Sometimes it was a difficult uphill battle. On some occasions I would probably have given up trying, but Paddy's quick wit and clear-headedness, plus long experience of dealing with 'super-clowns' always seemed to save the situation at the last minute.

One example of Paddy's unflappability was when an officer of 23 SAS, under heavy pressure to raise the number of men passing the selection course, decreed that instead of the selection course, the next batch of recruits would only fire a normal army range course to qualify for the SAS. I threw my usual wobbler which got me nowhere. Paddy just gave a knowing wink and said, 'Don't panic, Joe, we've got a couple of months yet. A lot of water will go under the bridge.'

Water went under the bridge all right, but every move to put things right stumbled into failure. Although Paddy and I thundered back and forth to the Rhayader area in mid-Wales and all points nasty every weekend, pushing the recruits through their normal pre-selection training, the hard fact hung in front of my nose. There was to be no selection course at the end of it!

Time passed. The date arrived. The recruits assembled and were kitted up as normal for selection course. I had great faith in Paddy but as the trucks roared out of Birmingham, heading for the rifle ranges near Brecon, I was wondering how the hell we were going to live this one down.

We arrived at the ranges and found another unit already in residence. There was a lot of waffling telephone calls and angry officers, but we found we had not been booked on the ranges. They were fully booked for the next two weeks.

Paddy came to the rescue. 'Okay, there's no problem. I can get accommodation in Dering Lines (Brecon). We'll just have to go walkies for a couple of weeks.' And so we did. It was obvious to me that the accommodation in Dering Lines, plus land clearance for all the selection course routes, plus a lot of other things like rations, had all been organised long before. Paddy gave a big wink. 'I must have forgotten to remind the Adjutant to book the ranges. Never thought to check on it.'

Selection courses for 23 SAS went on as usual after that. They were very hard and the pass rate was the same as for 22 SAS, ten per cent. There were always pressures to get more men through the selection process by lowering the standards, but thanks to 'the Nuge' and a few others, as far as I know 23 SAS has remained a well-selected SAS unit.

At some stage during my two-year stint with 23 SAS, some comedian with a vast sense of humour sent me on a parachute-jump instructors' course. Of all people to get stuck with such a thing. I hated bloody parachuting and everyone knew it. The PX parachute saved me; it was on its final trials before being issued to replace the old X-Type. Being a bigger chute it gave a slower rate of descent, so I survived the sixteen or seventeen jumps necessary to complete the course.

The course was run by No. 1 Parachute School, RAF Abingdon. When writing my report at the end of the course the RAF boss of the Parachute School put one of the most apt footnotes I've ever seen. It was a very good report and I had passed but, as a little footnote at the bottom, he wrote, 'Not suited to parachuting – either in size or inclination.'

As a measure of the very low profile favoured by the SAS in those days, there were a few words from a farmer's wife in the Cotswolds that I'll never forget. A big escape-and-evasion exercise was being laid on in the north Cotswolds and letters were sent to all farmers and landowners requesting permission for troops to cross their land. Some of us were sent out to collect the written permission and check for any problems.

Arriving at one farm I was greeted with a cheery 'good morning' by the good lady, who looked duly impressed by my smart appearance – SAS beret and badge, camouflage smock with black belt, olive-green trousers, shining black anklets and boots.

I wished her good morning, then said, 'I'm from 23 Special Air Service Regiment; we sent you a letter a couple of weeks ago requesting permission to cross your land.'

She looked rather puzzled, then it dawned. 'Oh', she said, 'you've come to spray the crops.'

There were many interesting and comical experiences during that two years, far too many to record as usual, but one in particular is too good to miss out. 23 SAS often do exercises for a few days in Europe, sometimes parachuting into some country during the dark hours, sometimes landing the normal way and being put in by road or helicopter.

We had parachuted into Denmark and were on our way home on Sunday afternoon after a fairly strenuous few days. Our transport was two ancient Beverlys, one of which developed a 'hiccup' in one engine. Both aircraft landed somewhere in a hayfield. I don't know in what country; must have been Denmark, Germany or Holland. One of the flight engineers borrowed a ladder from a nearby haystack, climbed up to the offending engine and routed out a lot of odds and ends, like nuts and bolts.

The aircrews decided the aircraft was okay to go again, but as it was a bit suspect it was decided all the men who had to be back at work on Monday should go in the other aircraft and all the regular army personnel and self-employed types go in the dodgy one.

We got everyone shifted around according to their work problems and both aircraft took off again. About ten minutes later the same engine (port inner) started playing up again and gradually got worse. I was going around the people in the freight bay, getting their names for the flight manifest because of the passenger change.

There were two young lads sitting together looking rather nervous, as well they might considering the noise from the ailing engine. I had just got the name and number from the first one when there was a decent-sized explosion from port inner and sheets of flame were passing the window they were facing. I couldn't resist it, I continued calmly with the next question, 'Next of kin?' They both looked at me in horror.

The pilot asked on the intercom if we had any parachutes on board. We had a few decidedly second-hand ones in a hamper, which no one was about to use, so we said, 'No.' He decided to head for the Dutch coast but warned he might have to ditch in the sea. We all got to ditching stations. Then the pilot decided he would make the coast. He had thrown it about a bit, presumably in an effort to put out the fire. From where I was sitting it had obviously failed!

The engine was screaming fit to bust, so I presume the crew couldn't shut it down. We reached the coast, with flames everywhere on the port side. The pilot decided our best bet was to make for Schipol, Amsterdam International Airport. So we flew on for a while, with great faith in the RAF, if nothing much else going for us. As we landed at Schipol the airport emergency services were with us before we touched down. Matching their speed to that of the aircraft they smothered the engine with foam.

We opened a door, put the little ladder in place, gathered our belongings and wandered off in an orderly fashion. The Dutch were amazed and tried to organise a panicky run. They said there could be an explosion any minute. One of the TA types said, 'I don't think so, you've all done a marvellous job. Just keep it like that till I've had a smoke.'

The engine was still smoking, hissing and glowing through its ever-increasing layers of foam, but it did no more than that and eventually subsided, as all good engines do.

In 1963 I tried to return to 22 SAS as the Borneo problem had begun and the regiment was involved in real operations again. I had all my arguments ready when I went before the CO of 23 SAS, but I'm not much good at army politics. The CO was a past master. He shot down all my arguments (to his satisfaction if not mine) and I came out of his office a firm, if reluctant, member of 23 SAS.

From then on I just kicked my heels and waited for the few months to pass. I was due to return to the fold in mid-1964 so patience was the order of the day. Easier said than done when your friends are being killed and you can't even get off your backside.

13 CLOWNS AND CLANGERS

Early in 1964, while still with 23 SAS, I had been sent back to Hereford for a few weeks to help staff the overloaded selection courses which were made necessary by the reforming of 'B' Squadron. 'B' Squadron being reformed was great news to all of us, as this meant the regiment would be more able to cover all its commitments without burning out the troops in a continual yo-yo of operations.

For a time, just before returning to 22 SAS, I had a worry that I might find myself being posted to 'B' Squadron. Nothing terrible in that really, but, although many of my friends were now in 'B', I felt a strong identity with 'D' Squadron, and had no wish to change,

In the event there were no problems. I returned to 18 Troop in mid-1964, felt immediately back at home and became submerged in a deluge of training. The training was fast and furious. Our feet hardly touched the ground before we were off on another caper. Any time we spent in Hereford we utilised for range work, mostly on the twenty-five-metre range near the camp, sometimes at the Ross-on-Wye rifle range, occasionally on the field firing ranges at Sennybridge, south Wales.

Every week while in base, I took 18 Troop to the Black Mountains for a bit of Bergen humping. The only way to keep fit to carry a Bergen, is to carry a Bergen. Gymnastics and jogging are great for gymnasts and joggers, but I had seen too many PT instructors from the rest of the army, loaded up with muscle and gymnastic ability, collapse in the first few days of a selection course. My theory was they had never been really knackered before in their lives, whereas being smashed out of our minds with fatigue was the norm for us lesser mortals.

On our trips to the Black Mountains we carried only the essentials for survival in our packs, topping up the weight to fifty or sixty pounds with rocks picked up at the drop-off point. The advantage of this was that if anyone had a slight injury he could dump the rocks and be left with a relatively light load while making his way to the pick-up point or one of the emergency RVs. A couple of times men arrived at the RV with empty packs. They didn't try to hide the fact and, as far as the rest of us were concerned, they had done the sensible thing.

We always dropped off in pairs from a truck on the west side of the mountains and made for either one troop RV or, occasionally, separate RVs along the road east of the mountains. The inevitable wet, cold ride back to Hereford was much shorter from east of the mountains.

A man's physical and mental ability to pull his weight, do his job and remain operational was largely his own responsibility. That is where the self-discipline comes in. The troop were never supervised with loading their packs for training walks. For all I knew I could have been the only one carrying any rocks, but it was in everyone's interest to make the most of these opportunities to keep their physical and mental abilities at a peak. In fact I knew some of them were carrying far more weight than me.

I always carried binoculars on my trips to the mountains, and if it happened to be clear enough to see I kept an eye on the progress of other pairs anywhere I got the chance. The reason was safety. A man is easily injured carrying heavy weights over rough country – even when he is an expert! Other members of the troop also carried binoculars for the same reason. Binoculars can save a lot of time and leg work when searching for anyone missing in the mountains. They can also let you see at a glance whether a man is carrying any weight or just an empty pack.

During my SAS soldiering years I had my fair share of good and bad officers. All the squadron commanders I served under were very good soldiers; men who could cope with responsibilities and situations far beyond the normal for their rank in the army. Field commanders who seemed to have a natural touch for knowing the capabilities of the men under their command, so that we could achieve the impossible, but miracles were never actually demanded (only tried for!). Troop commanders on the other hand were a different kettle of fish. Mostly they were too young and inexperienced to be expected to cope with handling an SAS troop on operations.

The good ones tried not to get in the way, learnt fast from their experience and finished their two or three years with SAS very much advanced in their knowledge of field work and just about able to cope with their position.

There were some, like Rory Walker, who were brilliant and great to work with. They were the type who could give and take, listen to the more experienced men around them and make the right decisions when it came to the crunch. But there were others.

Once, on a hoof over the Black Mountains, I had the misfortune to be accompanied by one of the others. As was normal practice in the troop, we dropped off in pairs along a road west of the mountains, making for an RV on the east side.

I was paired with 'his nibs'. We duly loaded our packs and set off.

About two-thirds of the way up the first climb he was way ahead of me, going well. Before the top he started shouting for me to hurry. At the top of the first climb he ordered me to dump my rocks, saying I was obviously too old for it and couldn't keep up the pace. I refused, so he got angry and told me he had dumped his rocks way back so I should do the same or we would not be first into the RV and, as an officer, he must be first!

I couldn't believe my ears. Was this an SAS officer? I asked him what would he throw away first on operations, his rations or his ammunition? He rushed on ahead, cursing me for my age, my lack of fitness and for existing. I plodded on as one does plod on with a bit of weight to carry, ignoring the patter.

At noon, finding myself in a sheltered spot with a good mountain stream nearby, I stopped to make a brew of tea to wash down the cookhouse sandwiches. The clown nearly had a fit. He rushed around here and there, cursing loud and long. I told him to go on ahead to the RV if he wanted, but he wouldn't. It occurred to me the bum couldn't trust himself to find the RV. But surely he had passed a selection course?

Luckily, that type was a rarity which never lasted long in the regiment, but I did have another, more difficult, problem.

We were training overseas and had a clown along who just had to do everything wrong. A lot of problems were caused, some annoying, others downright dangerous. Eventually I blew my top, a stupid thing to do, but then, I think everyone has his boiling point. Show me a senior SAS soldier who has not blown his top over clown problems and I'll show you a saint.

The training was nearing its end. 'Old Joe' (much younger than me) had a poisoned foot which eventually had to be lanced. He had struggled along on two hard night marches in some considerable pain, but after the lancing job he couldn't force his boot on, so was virtually off the road.

We were going off on light-order patrols for the day, leaving our packs at our last stop, near a village. Joe had been ordered to take his patrol out but no one had been detailed to stay behind to look after the packs, so I told Joe to stay put, one of the troopers could take his patrol, there was no problem.

It was a very hard day's work in my case, and I returned exhausted just before sundown, to be met by Joe who was very upset. The clown we had along had returned just before me, realised Joe had not taken his patrol, and tore into Joe, calling him a skiver, waste of rations and a lot of other things.

My natural reaction to this news was as natural as anyone's. I said, quite loudly, 'The bloody idiot!'

The clown stepped from behind a rock, yelling about being called a 'bloody idiot', ranting and raving in great order. When he eventually subsided it was my turn. I gave him a quick but precise rundown on his ancestral history, then brought him right up to date. For the first time ever he was speechless. Then he looked around for witnesses. Where there had been most of a troop there was now an empty space. Silence. Nothing. Telling him to go away in my best trog's English I got on with the important task of making a brew.

A few days later the clown tried to frame me and even had me under open arrest at one stage. Needless to say he was not the type for the SAS and didn't last long. On another occasion, in the middle of an operation, we were sent a new troop commander. I don't know his background, but I expect we were victims of the system whereby 'little snots' (young second lieutenants) on passing selection and further training were promoted to captain and made troop commanders when they had nary a clue.

Unlike the troggery who, to join the SAS, throw away rank, pay, promotion prospects and cushy jobs, to get away from bullshit and be real soldiers, the officers were automatically promoted to captain, as the SAS had no commissioned rank lower than that. This did their future prospects a power of good, often being a good boost up the promotion ladder. Perhaps, in the machinery of getting good squadron commanders, it was all very necessary, but it meant a pig's life on ops when a troop got stuck with a little snot.

This particular child tried to impress us with his great superior knowledge of all things military. He was a great success as a bloody idiot. One of his greatest cock-ups in the man-management department made him (and us) the laughing stock of the whole squadron. At the time we were a very experienced troop and were operating smoothly in the situation. The clown gathered us together and complained we were making too much noise when moving at night, then gave us a quick lecture on 'individual movement in the field' – followed by demonstrations of the 'monkey run', 'leopard crawl' and 'ghost walk' (very basic methods of night movement taught to every soldier during his first eight weeks in the army, mostly only of any use in European-type terrain – which we were most definitely not in at the time). At first, as he had only joined us about three days earlier, we thought he had a great sense of humour and was trying to relieve what he thought was boredom. Then we realised he wasn't enjoying his demonstrations as much as we were. Especially when, red faced and totally pissed off, he wanted us to practise the 'techniques' so he could check us for faults!

Most of us had at least ten years' military service behind us, a lot of it on operations. Several had been infantry instructors and for this wet-behind-the-ears prat with all of eighteen months' service to lecture us, it was unbelievable.

A few years ago I noticed the poor old Marines had the same problem. They ran an officer-recruiting advert in various newspapers and magazines. It had a group photograph of a reasonably hard-looking crowd of 'Royals'. The words under the photo read something like, 'Are you aged eighteen to twenty-five? Have you so many O Levels, A Levels? Could you lead these men?'

Anyone who answered that advert on its face value was a prime candidate for the funny farm, and certainly could not have come within the hat-size range of the Marines or any other service.

The officer, even on SAS operations, always has the advantage afforded by the chain of command. He gets the orders for his troop, then passes on what he thinks is necessary. This can often be a problem when he's a flaming idiot to start with. Information-wise, we were treated like mushrooms – kept in the dark and fed on shit! I am often assured things have changed considerably since I left the army but, having noticed very little change in officer problems since (according to history) the Charge of the Light Brigade, up to the time I left the service, perhaps I can be forgiven a little scepticism.

In November 1964 'D' Squadron went to the Aden Protectorate for a training stint. We hoofed around a lot of desert and mountains over quite a large area, west and north-west of Aden. A lot of operational techniques were tested. Some worked – some didn't. We did a few new-fangled things involving aircraft and a lot of hard, hard flogging. We spent five weeks in Aden, then returned to the UK, where we went into a bout of intensive training, punctuated by Christmas and New Year, preparing for our next trick.

After a bit of leave and reorganising we were off to the Far East. A war is a war, but the war in Borneo had to be called a confrontation. Presumably because we didn't flatten Jakarta and they didn't attack Malaya or Singapore with aircraft. The confrontation was more or less confined to one area: northern Borneo. Mostly only land forces were used to any extent in an aggressive role by either side.

In 1963, Malaya, Sarawak and Sabah, all ex-British territories, had joined together to form the Federation of Malaysia. Sarawak and Sabah, together being about the same size as mainland Britain, but almost entirely covered by dense jungle, have a seven-hundred-mile border with Indonesia and are on the north coast of Borneo, which is the world's third largest island. The vast bulk of the island is Indonesian territory, called Kalimantan.

President Sukarno of Indonesia had ambitions of making a federation with Malaysia and the northern Borneo territories, to add undoubted riches and roughly eight million people to his already vast empire of thousands of islands with a population of about one hundred million.

The Indonesian Army was large, well equipped and experienced in jungle warfare. Sukarno was becoming a little unpopular in some quarters of his own land, being a virtual dictator, so, when the Indonesian communists became entangled in a small revolt in the northern Borneo territories, and tried to exploit the situation, Sukarno decided this was his great chance. Units of the Indonesian Army were deployed in late-1963 and our forces in Sarawak and Sabah found themselves coming up against more and more regular Indonesian Army units instead of the original communist raiders and infiltrators.

Owing to the rugged terrain of Borneo, the border was generally a very isolated area, far from any lines of communication, making it very difficult for either side to deploy large forces across it. All movement was on foot as no vehicle could get anywhere near the border. On the other hand, because of the all-enveloping dense jungle, no movement could be seen from the air. So the Borneo confrontation was very much a man-to-man war. Rifles and bayonets at very close quarters. Sudden, savage encounters, with victory or defeat often obscured initially by the tangled mass of dense undergrowth. Leading scouts walked on a razor's edge, only the fastest gun had any future.

Due to the terrain and dense jungle, all movement of troops was, of necessity, in single file. So, when coming face to face with enemy forces, it was usually only the first man, the leading scout, who actually saw the opposition and opened the fire-fight. Neither side would have any idea how many enemy they were facing. It could be two or three – or two or three hundred. Also, the initial uproar of any contact is magnified and very confusing to ears which, in the comparative silence of the jungle, have been straining to hear the slightest noise.

As we only operated in small numbers and the opposition were normally anything from twenty- to one-hundred strong, it was not in our best interest to hang around until they could deploy their main strength, but it was crucial that we win the initial contact so that a clean fast break could be made. With a lot of good training and experience under our belts we were confident of our ability to win, but there is no accounting for the whims of Lady Luck and anyone can be unlucky. Every moment on jungle operations we had the prospect of being blown away. There are no safe places in the jungle.

It is quite obvious that unexpected contact with enemy forces was to be avoided in the interests of successful operations (to name but one reason), but this could never be guaranteed – even by some of us who were favoured by Lady Luck and almost fly enough to walk upside down on the ceiling.

At the start of the trouble, SAS personnel had been deployed in ones, twos, threes and fours along most of the seven-hundred-mile front to make friends and allies of the jungle tribes. The tribesmen have always been very pro-British and, although their liking for headhunting was often emphasised by grisly relics hanging in their longhouse, the SAS trogs cheerfully accepted these little quirks of behaviour and soon became almost part of their local tribe. The eyes and ears of the jungle tribesmen augmented those of their SAS guests, who passed on information of enemy movement by radio.

British and Malaysian forces deployed in Borneo had no chance of holding a normal front line to defend Malaysian territory, so, as Indonesian incursions were detected, forces were moved to intercept them. Sometimes the Indonesian attackers were followed for days by SAS personnel, in order to find out where they were going so that a right and proper reception could be prepared.

SAS personnel, like all other forces under Allied command in northern Borneo, were strictly forbidden to cross the border into Indonesian territory. However, by the end of 1964, everyone agreed this was giving the enemy too much scope to mount their operations. In consequence, SAS patrols were allowed to cross the border, on purely intelligence-gathering missions, to a depth of five kilometres.

The restrictions on our cross-border movements and behaviour on the other side, were gradually lifted as the war – I mean confrontation – progressed. Malaysian forces were almost all based in Malaya to defend against a possible Indonesian invasion. They were used to good effect when the Indonesians made an airborne landing in south-west Malaya.

The brunt of the fighting in Borneo fell upon the infantry battalions, notably the Gurkhas, whose jungle expertise kept one or other of their units at the forefront throughout the war. But it was eventually a Commonwealth effort, with New Zealand and Australian troops fighting alongside British, Gurkhas and Malaysians. All Commonwealth forces seem to work very well together, not only the front-line troops but also the vital supporting units. Although we were heavily outnumbered in Borneo, the skill and determination of those men in the forward areas prevented the enemy from becoming too big for his boots and the confrontation consequently turning into an extremely bloody war.

In January 1965 'D' Squadron again landed in Borneo. It was the squadron's third tour in that theatre, my first. We took over from the newly formed 'B' Squadron. Things had changed since 'D' Squadron had left Borneo in June 1964. The regiment was now committed to cross-border operations into Indonesia, in order to gain advance information of Indonesian attacks, and do a bit of tickling to slow down any build up of troops.

Squadron HQ was in a huge old mansion on the outskirts of Kuching. The building housed everything the squadron needed for prolonged operations: signal centre, armoury, ammunition, clothing and equipment stores, operations intelligence centre, briefing rooms, ration stores and all the clutter of odds and ends needed to keep a squadron on operations and in good shape; there was even a *char wallah*, complete with all his gear, in an outhouse.

To many people a *char wallah* would be a completely unknown character these days, but, to the British Army in far-flung places around the world, the *char wallah* was the saviour of the ill-paid troggery. *Char wallah* literally translated, means 'tea man'. But *char wallahs*, those turbanned gentlemen from India, up-to-date versions of Rudyard Kipling's water carrier Gunga Din, were the answer to all kinds of problems, like extra food and drink, laundry, sewing and mending and darning socks. They were even good for small loans – all at very reasonable rates. Some of them sold shirts, slacks, shoes, cameras, radios – anything for which there was a ready market within their unit. Some of them employed a dozen or so local people to do much of the work. I say 'their unit' because, although civilians, they were very pro-British Army and had pride in the unit they served. A good unit valued its *char wallah* and kept his problems to a minimum, both from within their unit and from without.

Our *char wallah* in Kuching was called Mac – short for McGregor (not a very Indian name!) – and he had been with us in Malaysia and elsewhere. His banjos (doorstep sandwiches) of egg, bacon, cheese or banana were often the first good food we encountered when returning from operations.

Squadron personnel were billeted in hotels in the town, mostly in the Palm Grove Hotel near the centre of Kuching. The Palm was quite big, comfortable and clean, with an ever-open and well-stocked bar. Right next door was an excellent Chinese eating house called Yan's Bar, so we were rather spoilt for comfort in the very little spare time we had between operations.

After two or three months we were knocked down a peg or two when we were moved to a large house on the outskirts of town and had our meals in a distant army camp. The situation was helped, however, by Mac, who placed one of his family (young Mac) in a room at the house and, once again, cheap food and drink was immediately available at any time.

The first day or two after arriving in Borneo we were thoroughly briefed on the operational situation, right up to date, then on expected enemy 'next moves'. We were told nothing about our own forces which was not essential for us to know. Most of the squadron was to be deployed in patrols, each of three or four men. There were four of us in my patrol.

We spent a short time polishing our jungle techniques and immediate action drills, then got down to work at last.

My first operation was a fiasco as far as I was concerned, although we brought back a lot of information which Brigade HQ found very interesting. Three patrols went in together, with one patrol acting as step up, while two patrols were tasked on to targets further on. The operation was scheduled to last fourteen days. Old Joe's patrol was doing step up; Alex's and my patrol were going further in. Joe's patrol could do some recce work if the climate felt right, otherwise he stayed put until we returned, then we would return as a three-patrol group through the border.

Being lifted into a jungle LZ by helicopter – rather than descending by parachute – may seem a rather tame way for the SAS to go to war but there were many reasons for it. The main one was that entry by helicopter avoided injury to us and to our weapons or some of our more sensitive radio equipment. Also, when we left a chopper on the ground we knew – and so did those in command – that we were intact as a patrol, immediately operational and exactly where we should be. Unlike a troop or squadron, a three- or four-man patrol is not the ideal formation to deal with casualties – especially the type with broken arms or legs dangling one hundred feet above the ground in enemy territory.

On this occasion, all three patrols moved a short distance from LZ, then stopped to get switched on. The noise of the helicopter had faded to nothing within a few seconds of lift off, to be replaced by the usual uproar of jungle insect noise.

We went south and west for about six or seven thousand yards, by which time we had found rivers where there were none on the maps and at least one big river flowing in the opposite direction to its course marked. It was mountainous country but no hills were shown on the maps. They were correct in one detail though: there was jungle.

Alex was in charge of the three-patrol group. At that time most of my jungle experience had been in Malaysia and other places where the maps had been quite accurate. Alex, on the other hand, had twenty times more jungle experience than me, and much of it with very doubtful maps.

When we reached the main river, very roughly where it was supposed to be, but flowing in the wrong direction, we got our heads together.

And we disagreed. I felt absolutely sure it was a tributary of the main river; so, if we followed it down for a couple of hundred yards it would curve around and join the river we wanted to cross. Alex decided the map was wrong right away (the intuition of experience) and said to cross it, keep on our compass bearing and to climb through thick bamboo and undergrowth up what was almost a cliff on the other side.

As leading scout I viewed the straight-on prospect with some apprehension. A lot of very steep uphill, covered in a tangle of vines and bamboo, compared to the relatively easy wander down the river-bank. Also, I felt certain we would be climbing that hill only to find we would have an equally nasty descent to the river the other side of it. In other words, a lot of unnecessary hard work.

I climbed the first five hundred feet feeling really pissed off. Then began to doubt myself as we went on climbing. Another five hundred feet and I knew I had been wrong and Alex was right. But it was a good lesson for me to start operations with, and get switched on to very dodgy maps.

The jungle in Borneo was very similar to the Malayan jungle in that it was normally very dense tropical rainforest. There were, however, a lot of things which made it feel different. Perhaps it was mostly the contrast in local geography between those areas I worked in the two countries. There were very rugged areas in Malaysia but, compared with most of the jungle areas in which I worked in Borneo, the Malayan jungle was rolling hills. In Borneo there were more cliffs, ravines, very narrow ridges and spurs. There were very few places in Malaysia where a distant view could be seen. In Borneo it was quite common to be able to see for miles across the tree tops.

In the leafy, cluttered twilight of deep jungle, fairly fast quiet movement was possible in most places due to animal tracks and old or little-used hunting tracks. Away from any type of track, movement was considerably slowed and usually unavoidably noisy.

When I say hunting or animal tracks, I'm referring to tracks which the untrained eye could never follow. The acquired knack for following such tracks, like looking through the trees instead of at them, takes some people a long time to learn. For me, such things were natural due to my childhood in the woods. The training I had received and my previous jungle experience sharpened my abilities in deep jungle so that I felt immediately at home, confident, on top of the situation.

To the untrained eye, the all-round view in deep jungle can be compared to that seen from the middle of an average room with two or three windows, at dusk. Except that the view through the windows would rarely be more than thirty yards. Walls of greenery blank out the remainder as solidly as the walls of your room. On the other hand, the trained eye will see usually at least twice as far all around but rarely more than forty yards anywhere. Most of the time even well-trained, experienced eyes can see no further than twenty-five to thirty yards. There are places where it is quite easy to see forty-odd yards in all directions. There are other places where it is impossible to see more than three or four yards in any direction. The absolute tangle of the jungle impedes all movement, but training and experience can always find a way through. There is no such thing as impenetrable jungle. The occasions when one is forced to use a knife or *parang* to make progress are extremely rare. Only once during the Borneo operations was I ever forced to use my knife to cut tangling vines, even then it was due more to anger, frustration and lack of patience.

One of the greatest problems of operational movement in jungle is the junk lying on the ground. Noisy, crackling dead twigs and branches often lie like a carpet. Dead leaves of all sizes and shapes rustle, crack and pop at the slightest touch. So it is of prime importance to take great care where you land your size twelves. There is almost always somewhere quiet to step – another acquired knack – made all the more difficult by the ever-present requirement to see your enemy before he sees you.

About six pairs of eyes would be very handy at times. Hanging vines and small saplings must be avoided like the plague. Just a gentle pull or knock at ground level can wave high branches like warning flags to signal your approach. Those hazards are mostly five or six yards apart at best in average jungle. Huge fallen trees and branches can usually be climbed over, or crawled under.

Sometimes it is necessary to make quite a detour to find a way past them. They occur much more often than the uninitiated might imagine, probably averaging one in every hundred yards.

Above about twelve feet there is usually a considerable thinning of the vegetation. Mostly treetrunks, the tops of young trees at various heights and the great rope-like vines trailing down from the jungle canopy, which again is very dense upwards from about one hundred and twenty feet.

Borneo lacked my pet hate: attap. More good news was that there were fewer leeches, at least I saw nowhere near as many as in similar areas of Malaysia.

We climbed a long spur on to a high ridge and, finding a small stream, made it our step-up RV. All was as it should be: no sign of enemy forces, nothing except the normal sounds, smells and sights of the jungle. We did an area recce for a day before leaving the step-up RV, so as to check the area for enemy signs and know enough to make the RV easy to find on our return. An hour away from the RV I realised I had lost my watch, which had been hung on a cord round my neck.

We were about five hours on that recce and I was moving back towards the RV, thinking I must be pretty close to it, when I saw fresh tracks a few feet ahead. The patrol stopped and I went forward to investigate. There was no doubt about it being our own tracks. The first thing I saw was my watch among the leaves.

The next day we moved off towards our targets. My target was a place called Kapoet, a small village believed to be an Indonesian Army base, possibly a Special Forces camp. My task was to get a look at the place and the troops in it, collecting all possible information about them.

The patrol had only reached a point a couple of thousand yards further than our area recce of the previous day when further progress to the west was stopped by a sheer cliff. We were confronted by a drop of roughly eight hundred feet. We searched the cliff top for about a thousand yards in each direction, but there was no possible way down.

The next day two of us went on a light-order patrol along the cliff top to the south, leaving the other two to look after the packs. We had covered about two kilometres when I heard a sound up ahead. We both decked and waited. The sound was approaching. It began to sound like something big crashing through the trees towards us.

Then I knew it was something big coming towards us. But what? It was certainly much too big to be human.

Elephants are rare in Borneo, but even they didn't usually crash about like that, unless they were terrified! Then, about one hundred yards ahead, I could see the tops of quite large trees swaying. I began to feel a bit nervous. Knowing the jungle, I knew it would take a tank to cause that kind of movement. There was no wind and in any case the movement wasn't wind induced. Every sense was straining to the limit to understand what was this monster bearing down upon us. I glanced back. Twenty yards behind me my wing man's eyes were like chapel hat pegs.

Then all was solved. Two huge ginger apes, one chasing the other, were swinging through the treetops. They came right over us, looking big and ugly but travelling very fast, swinging from branch to branch, tree to tree. The jungle canopy at that altitude was only about sixty or seventy feet above the ground, not as thick as that in the valleys, so we had a really good look at the apes, which I later found out to be orang-utans.

For several days we tried to get down the cliff. Searching a cliff top in dense jungle is much easier said than done. No animal tracks went more than about fifty feet down, but we had to explore each one. We could see the rooftops of our target a few thousand yards west, but there was no way we could get at it without a long, time-consuming detour. We were limited in our explorations to the western half of the plateau as we had no wish to bump into one of the other patrols somewhere unexpected.

Time ran out. We had achieved a big fat zero and returned to the step-up RV. Joe's patrol had been to check a jungle village, made contact with a jungle tribe and had great success while waiting for us to return. Alex had achieved more than I had in terms of distance but very little else. He too had been frustrated by cliffs, but had eventually found a route down to the south-east.

Before going in on that operation our packs had been weighed. We had been limited to fifty pounds for the fourteen days. The squadron commander, a damn good soldier well liked by everyone, made the fifty-pound rule with the very best of motives. Basically I agreed it was a good idea – but not in my case. A big engine tends to have a big fuel consumption, and I was no exception. Before the operation was over I was absolutely starving.

Luckily for me, Alex was trying to give up smoking. A fifty-pound pack gave him ample rations. So, when we all arrived at the step-up RV I carefully got upwind of Alex, lit a cigarette and let the smoke drift past his nose. After which it didn't take a diplomat to organise a bit of swapping so that Alex smoked and I ate. That was the last time I tried to operate from a fifty-pound Bergen. There were ways and means through the regulations, much easier than being half starved.

Returning to our base at Kuching I felt I had achieved very little, but at least our patrol was now well switched on to jungle operations. We faced future tasks knowing we could work well together. There being no suitable accommodation available in Kuching we were using the Palm Grove Hotel as sleeping quarters. It was in the centre of town and a good place for relaxing. We averaged four or five days between ops. The first day after returning was mostly taken up with a detailed debriefing by the SAS Operations staff and a session in the Brigade Intelligence Centre, where we passed on any information they required.

Brigade had a really good air-photo interpreter, a Staff-Sergeant Taylor, who helped me considerably with his superior knowledge of air photography. How he did it I'll never know, but from his air photos he could tell me the height of corn growing in a field to an accuracy of a few inches. And the photos were taken from a height in excess of ten thousand feet. I suppose that was the difference between an air-photo reader and an air-photo interpreter. Such information was vital when selecting a spot from which to watch the enemy, especially in a country where field crops could vary in height from two or three inches to six or nine feet.

Taylor also knew the rate of growth of any given crop and added on the appropriate inches according to the date of the photo. He made the planning of operations a lot easier than it might have been.

A day or two relaxing sometimes followed the debriefing, then would come the briefing for the next op. Every possible known fact about the areas, the enemy, the local people, the terrain and nearest friendly forces. Long studying of any maps of the target areas. Hours bent over a stereoscope studying every millimetre of air photographs. Practice of all IA drills with live ammunition. Test-firing weapons, then cleaning polishing and oiling them very carefully. We all carried self-loading rifles in my patrol.

The SAS were not the only residents of the Palm Grove. There were various other service personnel billeted there, among them some Royal Navy types who were with minesweepers operating around the Borneo coast. On a couple of occasions I noticed a very drunk, rough-looking individual in civvy clothes consorting with the Navy types. He was of stumpy build, always in uproarious spirits. I wondered where he fitted into the scheme of things. Somehow I knew he wasn't Navy. Didn't much like the look of him either.

One operation was another crack at Kapoet and its environs. For various reasons I had a scratch patrol; two of the patrol were men I didn't know. Again we went in with Alex and Joe, but this time dispensed with the step-up method, going our separate ways from an LZ right on the border. Kapoet is on the north bank of a fairly large river. On the north side of the river it was surrounded by cultivated land. Several different crops varying in height from a few inches to a few feet. On the south bank of the river there was a cultivated area stretching from opposite the village to the edge of the jungle and to the east. It was about four hundred yards square. West of that cultivation was young secondary jungle from right opposite the village to the primary jungle, about four hundred yards south. (Primary jungle is that which has not been cleared for at least one hundred years, if ever. Secondary jungle is that which has grown again where once it had been cleared.)

The difference is very apparent on the ground and also on air photographs. The reason for the difference is that most jungle trees grow at the same rate so, when they get a fresh start and all grow together they are all roughly the same size and usually grow thicker together on the ground. There are no giants among them to keep out the sunlight and no spaces left by dead or fallen trees. The trees become dense and uniform – very unlike primary jungle which, by comparison, is often a pleasure to walk through.

Putting myself in the Indonesian commander's position I would have expected trouble to come from the north, the direction of the border. I would not expect an enemy force to cross the river and approach from the south, where they would have to cross the river again to reach the village.

My safest approach to the target was therefore from the south, where we would have the cover of secondary jungle right to the river bank. That would put us within fifty yards of several buildings on the opposite side of the river. Closer approach from the south, however, would only be possible during the hours of darkness. This would depend on the state of the river, which at Kapoet is narrow and deep. No bridge showed on the air photos but, as the jungle completely overhung the river in places, there was a good chance of a rope bridge somewhere. Staff-Sergeant Taylor confirmed my opinion that the village could be seen from the primary jungle to the south-east, as the crops in that area were only eighteen to twenty-two inches high – or would be by the time I got there – and the ground rose gently up to the jungle's edge.

There were snags, of course. The river could rise and change completely in less than an hour, stopping us from crossing to the south side, or cutting us off for days once we were there. It was that sort of river. There could be, and very likely were, snags with that all too easy patch of secondary jungle. If, as our people thought, Kapoet was a Special Forces base it would most likely be guarded by more than a few mines and booby traps. Special Forces, if they were any good (and they were said to be very good), would certainly take precautions with that area so close to their base.

My task was as before: to carry out a recce which would make Kapoet a target for future offensive operations, or cross it off the list for a while. Top priority, as always, was the safety of the patrol; casualties must be avoided. I was forbidden to initiate any offensive action on this operation as the information was deemed top priority.

The four of us got our heads together over all the facts and figures we had collated. We estimated time and distance, danger areas, ways and means of overcoming this problem or that. We made a very flexible plan and argued the pros and cons. So, in we went.

After getting quickly off the LZ, the patrols worked things out their own way. Quickly but quietly, my patrol moved two hundred to three hundred yards away. Sometimes it was more, sometimes less. On finding a suitable place, I would give the signal they were expecting and the patrol would stop, take off their packs and adopt an all-round defence posture. We would sit and listen, using all our senses to get switched on to that particular area of jungle. No one would make even the tiniest movement for ten or fifteen minutes. Once satisfied that things were as they should be, I would take out my map and compass for a last check on our route and that what I had seen from the chopper, from the LZ and on our short walk, corresponded with the map and other information, so I knew we were in the right place. Taking out the map was the signal for the others to do final checks on their kit before moving off. There would be no words spoken and no talking again until we were on another chopper in a fortnight – or however long it was before we were lifted out. Everything was done by signals, or very low whispers when necessary.

The approach march was unremarkable. We saw no sign of the opposition and few signs of the jungle people. Reaching the target area we did an area recce of the jungle up to the edge of the cultivations, spending a lot of time moving slowly and listening a lot. No signs or sounds rewarded us.

On about the fifth or sixth day of the operation we had done a recce north looking for tracks, which took two or three hours. I then decided to cache the Bergens and move towards the village through the cultivations to find how close we could get on the north side. We didn't get far. We hit a patch of open grass which looked like a lawn. There was higher ground ahead and I started across, only to find there was a foot of black sloppy mud under the 'grass'. Seeing no quick way around it I went on through – leaving a huge track. Then found a well-used track, wide enough to drive a truck on, running along the higher ground. There was no way anyone on the track could miss seeing my tracks through the mud.

We pulled back to the edge of the jungle. I made a quick decision. We would compress the whole recce into what was left of that day – about five or six hours of daylight. A shit-or-bust move through the rest of our rough plan, ignoring the north approach, going straight round to the south side for a look-see. Trusting in our jungle expertise and, if it came to it, our immediate action drills to get us away with it.

At that point there was a tremor of doubt from one of the patrol but I ignored it. We headed for the river, ready for anything. The usual river track was big and on the north side. As with all other tracks we had found – no boot prints. Finding a good crossing point, we used our own patrol crossing method. This entailed watching the opposite bank for a couple of minutes then, with two to give me covering fire if required and one watching their backs, I moved out to the centre of the river. At this point on this particular river it was about seventy yards wide and just over my knees at the deepest.

Reaching the halfway point I turned and checked the riverbank I'd left, as far as I could see in either direction. If all was clear (and I didn't get shot), I moved on to the far side and checked a few yards around in the jungle. If that all felt okay I returned to the riverbank and signalled the next man to cross, covering the far bank while he did so. He would cross near me, then go a few yards into the jungle and cover my back while I waved the next one over, and so on.

I noticed a big fish trap in the river a few yards downstream. It was not in use. We had crossed the river about eight hundred yards east and upstream of Kapoet, so, once into dense jungle, we moved quickly in a westerly direction to hit the edge of the open cultivation south-east of the village. After about twenty minutes I could see the open ground through the trees, so changed direction slightly south so as to follow the jungle edge to higher ground, from where I knew I could get a good look at the target. Reaching a high point we moved carefully to the edge of the trees and there, as promised by the air photos, was Kapoet.

Half an hour was spent in that position, using binoculars to examine anything and everything which might tell us of the enemy presence. The range was about five hundred yards. Nothing stirred. There was no sign of anything. The joint was dead.

Rain began to fall as we moved on to the west. This rain was good news as it would cover the sound of faster movement. We made good time and reached the point where the young secondary jungle joined the primary. Now was the time for extreme caution. This was now the nasty bit. Every bit of ground was examined for signs of booby traps and mines, also every tree was checked before we passed. The patrol was on a hair trigger, moving very slowly. There was no sign of recent human presence anywhere, and in a place so close to a village that was bad news. The further we went through that mess of trees the more sure I became that something was wrong. Then we were through. The river track was a few feet in front of me. Beyond was the deep cutting made by the river, about thirty feet deep with almost sheer sides and twenty-five to thirty yards wide. Thatched roofs showed on the far bank, the nearest being less than forty yards from me. Leaving the patrol on stand-by I moved parallel with the track for a few yards, checking for bootprints. There was none. The track had been well used. Nothing much had moved on it for at least two or three weeks, probably a lot more.

There was a screen of young trees and bushes between the track and the river, the patrol came forward to deploy along the edge of the track so I could better do a binocular recce of the village.

Then our troubles began. The man who had been getting more and more nervous suddenly cracked. Under other circumstances I would have hit him, but if I hit too softly or too hard we would have a bigger problem. We had enough problems already. There was a job to be done and we were going to do it. Leaving the patrol in a situation of one holding down one, while the other kept watch, I moved off to do the recce. Their orders were to leave the way we had come (which was the only way out) if there was any shooting before I returned.

Now the rain became a menace. I would not hear anyone approaching, and that is not clever when you have your eyes glued to binoculars, or just looking in the wrong direction. I had to keep moving further along the track to see into buildings and around buildings on the other side of the river. The place looked deserted. No soldiers, no people, no animals or chickens, no clothing hanging to dry, no smoke, no steam. Nothing. I got right on the edge of the river but could see no bridge of any kind. Strange. Why no bridge? There wasn't even any sign of a bridge having been there in the past. No track ending at the river bank. Nothing.

I began to worry about the patrol. The binoculars were continually steaming up in the now heavily pouring rain, so I put them away. I looked hard at the river and its banks. If we crossed and I was wrong about the place, there would be little chance of getting back across to safety.

Being narrow at this point the river was deep and fast flowing. I didn't much fancy my chances of crossing safely. Odd jagged rocks breaking the swirling surface were decisive. No bloody way. The thoughts of fighting a war with that at my back was decidedly off-putting.

So what more could we do to make sure there was no enemy base here? I searched further west along the river, found nothing west of the village so returned to the patrol. I felt I could have spent more time looking in the right direction if I had been accompanied on the trip along the river, but it was impossible under the circumstances. Things had changed little, except the panic-stricken soldier was no longer being physically restrained.

Taking the bull by the horns we moved east along the river track to the edge of the cultivated area. Looking across the open ground I noticed an earth bank running from nearby to the edge of the jungle, right across the middle of the open area.

The bank was about five feet high, man-made, but was old and may have been to do with drainage. I checked it out and decided we would use it, partly to get away back into the jungle and partly to make a final check on the village.

At its nearest point the earth bank was about one hundred yards from, and in full view of, the village. If there were troops in the village they would be unlikely to resist shooting at an enemy soldier walking along that bank.

The light was very bad and the rain was falling like stair rods. The weapons of troops engaged in jungle warfare are rarely well zeroed, as long distance shooting doesn't happen, and the rigours of jungle movement are bad for weapon sights. Also I have very little faith in anyone's ability to hit me.

I put the plan to the patrol, two of which agreed and one of which went into shock. Then he refused to move. There was no way he could be coaxed to his feet, he sat and shivered, looking like terrified death. I thought back over the years to Lieutenant Peal of the Glosters and his instant cure for this type of shock, putting the muzzle of my rifle hard under the man's chin, I clicked the safety off loudly, told him he had three seconds, counted 'one, two' and he was on his feet. He knew as well as I did that I would have shot him.

I walked along the bank to the edge of the jungle, while the patrol kept their heads down and moved, well spread, level with me on the lea side. No shots, no problem, nothing. It was about three hundred yards, maybe four hundred yards, across open ground in full view of the village all the way to the treeline, where we were about four hundred yards from the village.

We watched for a few minutes, but there was no reaction. The village remained dead. Nothing moved anywhere. We had about an hour before dark, so moved off quickly through the jungle to try to reach our Bergen cache before the light failed.

I took a route which looped to keep well clear of our route coming in and hit the river after about twenty-five minutes. We crossed as before and I congratulated myself on my navigation as I could see the fish trap about one hundred yards away.

Just as the last man reached the jungle I saw movement some distance along the river track, to the east. Going into an immediate ambush position we watched as two jungle tribesmen came towards us. They were unarmed except for long *parangs*. The older man had his on his shoulder. As I stepped into the track a few feet in front of them I kept a sharp eye on that *parang*. The tribesmen looked a little nervous, but no more than expected: we had no time for the niceties of life and I thought the best move was to question them on the whereabouts of enemy forces right away, while they were a little off balance.

My problem was language. My patrol linguist was in shock but I called him forward anyway, hoping he could get some information. He stood staring at the two men, then blurted out, 'Apa matchum?' They stared back at him, as well they might. He repeated his question about four times before I sent him packing. I knew enough Malay to know his question was 'How's it going?' or words to that effect.

I gave the men the big grin, a thumbs up and waved them on their way. Before they were out of sight we were walking eastwards on the track, to put any would-be followers temporarily in the wrong direction. Then we skipped off the track at an appropriate spot, leaving no obvious tracks.

Before we left the river I noticed it was rising fast and felt good to have crossed it in time. The rain roared down like it was trying to bore holes in us. Knowing exactly where I was, I took off at a fast pace to find the Bergen cache.

Then things began to go wrong. The ground didn't go like it should. The jungle thinned and we were in a densely overgrown *ladang* (the cultivations surrounding Kapoet on the north side). It was impossible to see more than a few feet in any direction. I rushed on, smashing my way through a wall of greenery so thick that in places I had to throw all my weight forward and lie on it to make an impression. In places there were strong tangling vines which forced me to use my knife.

Eventually I stopped, exhausted, and sat down in the mud to reassess the situation. The patrol gathered round and we had a quick discussion. Everyone was baffled. Navigation-wise everything was wrong. We had a smoke. All of us thought back along our route. I tried to concentrate on the ground we had crossed after leaving the Bergens, before reaching the river. The others did the same. The only fact we could pin down was the fish trap in the river. That had been the point common to the route in and the route out, apart from the river itself.

I knew from my hours of studying and memorising the air photos there was no isolated cultivation surrounded by jungle. There was no chance Taylor and I could have mistaken this *ladang* for jungle, even though its growth was ten or twelve feet high.

The only thing which could have thrown me was that bloody fish trap. It wasn't the one we'd seen on the way in. Had it not been there I would have carried out the usual navigation procedures to establish our position before leaving the river. I studied the map and compass, working out the possibilities and impossibilities until I had something to go on, then off we went again, due east.

As darkness fell we reached the friendly arms of the jungle. Now I knew where I was to within a couple of hundred yards. The relief was great, but the Bergen cache was at least one thousand yards away through jungle which was bad enough in daylight, no chance at night. Except for the obvious exception, morale was high. I had no worries about the patrol right now, but I did think a lot about what shape we would be in tomorrow. I was pretty well whacked, as I deserved to be, but sleep was difficult in the ice-cold pouring rain. We had about eleven hours of darkness to look forward to, with no shelter except what we could rig with leaves in darkness. That wasn't much.

I deserved the lesson, the rest of the patrol didn't – but they didn't curse me too loud or too long. You can't win them all, but what a time to pick for bum navigation.

If there was a Special Forces base anywhere near we could expect a follow-up the next morning. On this type of operation, until we made contact with the enemy, or gave away our presence in an area, we were very much the hunters. But, due to the enemy's great numerical superiority, once they knew we were there, we became very much the hunted. They would quickly deploy considerable manpower with one thing in mind: our destruction or capture.

For our part, I'm sure almost all of us did a mental change from 'hunter' to hunted at the appropriate moment. Nearly always, on our return to the border (when we would head for the pass as we called it), we would definitely feel hunted. It was fairly certain the enemy would have discovered our tracks somewhere during the couple of weeks we had wandered among them.

I had plenty to think about through the long cold night. We had two brews of hot coffee during the night but ate nothing of our emergency rations. They were for emergencies, not bog-ups!

As soon as enough light filtered through the treetops we headed for the Bergen cache, and found it with little problem. My troubles, however, were not over. Being a person of habit, especially when those habits concerned staying alive (and were strongly recommended by the operational directives of the regiment), I took one man to check the area for some distance around before settling down to eat, drink and begin signalling. How I did it I don't know, but I got lost on that stupid little area recce and couldn't find the Bergens again. All kinds of excuses spring to mind but that's all they can be – excuses. And weak ones at that! It took us over an hour, and several animal calls, to locate our Bergen cache. My navigation was about as good as a drunk in a maze.

Reaching the Bergens at last, then moving to a new site, I told the patrol there would be no movement that day, a decision made during the night. We slapped an ambush on our track and played the waiting game. Anyone wanting to play bloody fools could come and get us. We were not going out to play until we were fit.

Next morning we were fit and commenced the march back to our LZ on the border. All went well until we reached the LZ. We had been told by radio that morning there were no friendly forces in the area, except the other two patrols who would be lifted out at the same time.

The patrol checked out the LZ area, then moved in to check the LZ itself. We were the first patrol to arrive, so posted ambush sentries and rigged the radio to report our arrival to base.

The signaller was working on his set at the west edge of the LZ, near a couple of native huts. As I couldn't leave the 'problem' on his own, he and the other man were in ambush on our back track. I had just been to check another track and was halfway back across the LZ right in the open, when a pair of little brown men in uniform, carrying pump guns, came walking around the corner of the nearest hut. They were about thirty yards from me and taken by surprise more than I was. I would have shot them, but just in time I realised the white bands around their jungle hats meant they were supposed to be on my side.

As this was happening I heard a slight sound to my right. Turning I saw the head and shoulders of Joe's leading scout coming up over the hill about forty yards away. The two little brown men saw him at the same time, and were in some consternation, as well they might be. I quickly called to the leading scout and smiled at the little brown men, letting my rifle hang down loosely in one hand. The tension eased, guns were lowered. A platoon of Scots Guards came wandering along the track behind their two little brown scouts.

The consternation of the wee men over Joe's leading scout was understandable. He was Fijian and, like many of our people, wore a sweat band – Geronimo fashion – rather than the usual jungle hat. I hate to think what would have happened if any one of the people there had been trigger happy, or if I had not been where I was – an obvious Brit. There would have been a war between us and the Scots Guards most likely, before anyone could recognise anyone else. The jungle is no place for slow reactions or the niceties of 'Who goes there?'

Alex's patrol arrived a few minutes later. We were all lifted out within the hour. Arriving back at base we bitched heavily about the signal stating no friendly forces in the area, which all three patrols had received that morning. I felt a great relief when we handed in our weapons after that operation. Relief that I no longer had a heavily armed potential nut case – who probably hated my guts – walking behind me.

The 'problem' assured me he would leave the regiment as soon as possible as he would play on a medical problem which would get him out of the army. He left as promised.

A lot more than is written here has some bearing on the 'problem'. I should mention he had been on operations previously where his patrol was in an extremely bad situation, entangled with enemy forces for several days.

It takes all sorts to make a world, an army, a regiment. There are some sorts an army and a regiment would be better off without, but a world? That's beyond my ken. The nervous strain of SAS-type jungle operations is well known to any who have participated in such activities. Throw in a few bad experiences at the wrong time and how many more would crack?

Talking of the all sorts to make whatever, this chapter reads like a catalogue of cock-ups – mostly mine!

14 BORNEO

My regular patrol was a man short and I thought it likely my next job would have to be done with a three-man patrol. This didn't bother me but I told my two remaining patrol members to think hard about our operating procedures and contact drills, as we might have to do a bit of revision. Most of 18 Troop had different procedures and contact drills to those as ordered by the squadron and regiment. We had a lot of hard, long experience in the troop and pooled our brainpower in all things operational.

Every time we went on the jungle range between ops, we had to practise the contact drills as laid down by the squadron, but by various means I made sure my patrol practised our own contact drills as well. It was our lives that depended on them.

Then Paddy, my signaller, was rushed into hospital for an appendix operation. That left two of us: Pete and me. Right at that time the squadron was really pushed. An Australian infantry battalion was taking over from the Gurkhas on a notoriously bad stretch of the border. An Indonesian attack was expected at the worst moment, before the Aussies had time to get to know the ground. To give the Aussies some warning and help break up the expected attack, the squadron was to be deployed over the border a few thousand yards, as a screen along the battalion front.

A spare signaller was found to take Paddy's place and an old friend, Bert, came as my medic. The unstoppable Bert, whom I had first met in Korea, then at Brecon, had been a member of 'D' Squadron for a long time. We had done a few jobs together before. It always amazed me to realise Bert had been in the Anzio landings during the Second World War. Now here he was again, twenty-odd years later, still soldiering on – and doing a damn good job of it – in one of the most demanding types of military operations. What a player! What a bloke!

We were genned up for the job, then spent some time working on our contact drills and operational procedures. There was little problem with maps and air photos as we only had to reach our designated position, then lie in wait.

The patrol was flown to the forward positions, then escorted by a

Gurkha platoon to about a thousand yards short of the border. The border at that point was a small, sluggish river, flowing through sparse jungle, scrub and *ladangs*. The patrol approached the border cautiously, for Indonesian patrols were known to be very active beyond the river. Slipping into the cool, murky water I eased slowly across. The water came up to my armpits at the middle point but there was hardly any current. I did the usual check around on the far side, then, finding no sign of problems, returned to the riverbank and signalled the next man to cross.

Bert was last man, and not much over five feet. As he reached the middle of the river, I curled up. There was just a hat, two hands and a rifle showing above the water. Bert reached the bank, ejected a mouthful of water (no fish!), whispered, 'You-long-bastard!' Then he slithered out of the water and into the undergrowth.

For various reasons we had to make a radio call to base as soon as possible after crossing the border on that particular job so, about one hundred yards from the river, I stopped and gave the signaller the go-ahead to send the appropriate codewords. The radio sets in use at that time had no voice capability, all signals were in morse code. A wire aerial also had to be strung in the trees in the right direction.

Bert went about twenty-five yards one way from the signaller. Pete went about the same distance in the opposite direction to watch and listen for any problems. There was nothing hard and fast about their positions. Trained men instinctively find the right place at the right time. I had strung the aerial and went back to the signaller to check it was working okay. He was sending our call sign, then listening for an answer. I heard the morse screaming in the headphones and glanced at Bert. He was standing with one foot on a log, his rifle across his knee. I knew something was wrong; Bert was motionless, frozen – but coiled.

Then, under some branches about fifteen feet in front of me I saw feet moving. An Indonesian patrol was passing us, going towards the river. I glanced down at the signaller who had just turned down the volume of the radio.

Our glances met and he knew. He froze. Sitting with his back to a tree, not big enough to hide him, the headphones over his ears, unable to move he stared at the rifle lying at his side while the sweat poured down his face. He could see nothing, so knew the problem was behind him.

Very slowly I turned my head to look for Pete. He was not where he should have been. I couldn't see him. Shit! Where the hell was he? He wouldn't have gone for a shit without telling me. Was he aware of the enemy, or were they about to bump into him? The signaller was now watching me, standing about three feet in front of him. I knew he was waiting for the first signs of trouble to go for his rifle. Bert was still frozen.

I hadn't been able to count the feet passing, but the only ones I had seen had been bare feet, no army boots. I knew it wouldn't be local tribesmen, for they had more sense than to wander about in an area shattered by shell fire, where men hunted each other in the jungle and tall grass. The Indonesians, like us, employed local tribesmen as scouts and trackers, to move ahead of their troops or with them, as the case may be.

We waited a long time, motionless, until all was quiet. Then Bert was looking at me. I looked at the signaller, nodded towards his rifle. Thankfully he took off the headphones and slowly picked up the rifle. I signalled Bert to stay put and turned to go and look for Pete. He was stood right behind me. I said, 'What the hell are you doing?'

He said, 'Well, you always told me – if there's going to be any shit flying – get behind something thick!'

What can you say? In fact he had good reason to be where he was, but couldn't resist the funny stuff.

We moved off quickly and sent our signal from another place. But the worry was with us that the opposition might find the unavoidable marks where we had crossed the river. Good trackers wouldn't have much problem tracking us from there. The usual precautions were taken but the terrain was such that tracks were inevitable in places.

Arriving in our appointed patch, a swamp, which luckily was dried up at that time of year, we found a good spot from which to carry out our task. There was one problem: water. We didn't want to make tracks looking for a stream. We had seen no water since leaving the border river, but water we must have if we were going to stay there for two weeks.

Bert got to work on the problem, digging a hole in a likely spot. After about twenty minutes there was enough black, oily-looking water in the hole to make a brew of tea. Bert then went to work on straining and sterilising the liquid to remove the impurities – obvious and otherwise.

Pete became interested in the techniques of using socks, towels and sand to strain the water. One thing led to another. Before I knew it, Pete was being taught all the ins and outs of the medic pack. We were on shift, two at a time, and every spare minute between shifts Bert and Pete were in a huddle. The Indonesians saw fit to steer well clear of us, so, at the end of two weeks, Pete had received a concentrated medical course. On Bert's recommendation he did the patrol medic job in my patrol from then on. Pete had, of course, already done an SAS medical course before joining the squadron, and had spent time on hospital attachments working in casualty departments in the UK. Nevertheless, before a medic would be entrusted with the responsibility of being with a patrol on operations, they required more than a few good passes on paper and experience under clinical conditions; what they really needed was experience of SAS operations.

The time came to return to friendly territory, by now held by the Australians. When we met the Aussies they were quite surprised. Their people had written us off due to two of their patrols taking a fix on a short sharp battle and coming up with our location as the hot spot. We had heard several skirmishes and some artillery shelling at times, but assured the Aussies nothing had happened in our patch.

After partaking of some really good Aussie hospitality, especially in the food line, we were given a lift in a chopper back to Kuching. Base were glad to see us as they had been a bit worried by the Aussie reports of a contact being heard in our patch, even though our radio reports had been on time. They couldn't understand us not reporting enemy activity in our area. There had been a suspicion the signaller may have been captured and forced to send negative reports. It was all a bit upmarket from reality for us, so we handed in our rather humdrum report and departed to drown our sorrows.

Later, the squadron commander said to Bert and me that he'd had no worries we'd get ourselves captured again. I wasn't sure if this was a separate joke or a continuation of the one he'd pulled before we went in, when he said, 'Hey, you ex-Glosters, make sure you've got no white handkerchiefs in your pockets on this one!'

The spare signaller who took Paddy's place on that operation was, as well as being a very good signaller, very prone to letting his hair get rather long. It was thick and blond and very noticeable when beyond the bounds of military good sense. As he had been in my patrol the powers that be leaned on me to get his hair under control. I told him to at least go for an estimate soon, just as a hint, but next day the heavy thatch was still there.

The way we were living at the time there were no secrets. The hair problem became a prime target for the comedians, and they made the most of it. The first I knew about it, Pete told me in a whisper, 'You were a barber in civvy street – one of the best!' A few minutes later the target was being wound up on how he could save himself a trip into town and the price of a haircut if he could talk me into giving him an expert trim.

I did my bit by playing hard to get, claiming I'd joined the army to get away from barber shops, so I didn't want to know. It worked like clockwork. Within minutes I found myself standing over the mop with scissors and comb at the ready. Lots of snapping of the scissors and flexing of wrists, then I got stuck into it.

The 'idiot handle' (as it's known in my trade) was soon lying around in tufts all over the place. I made such a mess he had to go straight into town to get it tidied up – or at least down – almost to the wood! How the comedians and the rest of the crowd kept quiet during the cutting was a miracle, but they made up for it afterwards.

I reported to HQ to pick up the details and briefing for my next op. Having got the general details I was called by the squadron commander. He had a chat to me about my next job. We discussed the pros and cons for a few minutes, then he glanced past me towards the door, and said, 'Ah, yes, Lofty, here's your replacement.' I turned to the doorway and nearly asked him if he was joking. Standing in the doorway, looking like he always did, was that stumpy, ever uproarious, always inebriated character I'd seen a few times in the Palm Grove.

I shook hands in a daze, but noticed the twinkling blue eyes were now stone cold sober. There was hope. Little did I realise at the time I was looking at the final part of the jigsaw that made a team complete. This was Kevin, a Yorkshire lad who (I soon found out from the Ops Room) had gone on three operations but only spent four nights in the jungle, having been chased out twice by the Indonesians and shelled out once by our own artillery. He had a knee injury which was still suspect, but was keen to get into the jungle again.

That's all I needed, a dose of bad news – and a suspect knee thrown in for good measure. I knew we were pushed for men, but why pick on me? A three-man patrol looked a far better proposition.

Later that day I introduced Kevin to the tried and trusted members of the patrol. Pete, who may have been a barrow boy in London (rumour had it), couldn't treat anything or anyone seriously, made a joke out of everything and everyone, a born comedian. Paddy, from southern Ireland. Scruffy as hell unless he wanted to look smart. One of the best signallers in the SAS. Treated his radio set like the crown jewels and everything else like shit – except his rifle! I always swore he only looked after his rifle so he could look after that bloody radio. Another comic, whose occasional seriousness was ridiculed into laughter by Pete. Out of the jungle they forever clowned and got into trouble. On operations they would have a go at anything. Their coolness and guts could only be admired. But I was the only one to see them on ops, when it counted. Alex summed it up one day. He said, 'Lofty, I wouldn't go in the jungle with any one of your comedians – let alone all three!'

The first operation Kevin came on I put him where he would have to be if he stayed with the patrol, at the back, 'Tailend Charlie', a position which has its special responsibilities on that type of operation, calling for an 'over the shoulder' alertness. We had hammered him with our own brand of operational procedures and contact drills. He took to it like a duck to water. (An unfortunate turn of phrase in Kevin's case!)

Although Paddy had lost his appendix in a hurry less than three weeks before, he insisted he was fit enough to be back in the jungle. I warned him, if he gave me any trouble he would get the three-seconds-to-move treatment – but with the rifle barrel in a more painful place, out of sight! He in turn claimed he couldn't trust us to go wandering in the woods without him. So, after checking with the medical types, I gladly accepted Paddy and his undoubted expertise back into the patrol.

Again we were to look for the enemy Special Forces base. I looked forward to another fourteen days of very dodgy movement, looking for trip wires, booby traps, mines and the like. As Kevin said, we couldn't possibly go on getting away with it, never getting chased out – especially with him along!

The patrol had not covered two thousand yards – we had just reached the bottom of the border ridge, in what was considered a hot area – when I came to a small stream, about six or seven feet wide. I jumped the stream easily, as the far bank was a bit lower than the near one, then moved on along the bottom of a cliff. The patrol were all in place behind me as I rounded a corner of the rock face, so I kept going. After about thirty yards I stopped. Pete had not come round the corner. I waited. He didn't appear so I cursed and started back.

As I rounded the corner I could hear a splashing sound. A few yards away Pete was in a state of collapse, tears of laughter streaming down his face. I walked past him, not seeing the funny side, to find Paddy sitting on the ground, choking with suppressed laughter. The splashing was louder. Then I saw it.

Kevin – being stumpy – had got right to the edge of the higher bank of the stream for his jump. The soil had given way and his feet had gone through into some tree roots. He had fallen forward, face down in the stream and the roots had a perfect lock on his feet. He was doing press-ups to breathe as his arms weren't long enough to keep his head above water.

I had to kick their backsides to go and get him out. I think a bit longer and he would have drowned. The whole scene was like the Keystone Cops. Kevin was savage as hell when Paddy got him out, but within a few minutes was curled up with laughter at their descriptions of his predicament. There was nothing wrong with his sense of humour. But I wondered about the suspect knee. Kevin declared himself fit and off we went again. The knee was obviously no problem.

From that time on I realised that if we were all killed on operations it would be most likely we would all die laughing. The worse things got the harder those three threw themselves into it. Morale was never a problem. The only problem I had was suppressing the high spirits when the situation demanded.

Communication between patrol members was all in signs and very low whispers, but that didn't stop those three from grinning.

On one of our early operations in Borneo I did manage to stop Paddy from grinning. As mentioned earlier, he was rather careless and scruffy with his kit – apart from the radio, its equipment and his rifle. Most of the kit in our Bergen rucksacks was packed in waterproof bags, so a bit of damp getting in the pack didn't do much harm. In Paddy's case, however, he usually managed to have to search for some item before we had been in more than a day or two. When Paddy searched his pack he would often pull out waterproof bags, and empty their contents back into the Bergen. We called the result 'Bergen Pie'. To compound this problem, Paddy also had a bad habit of forgetting to close and fasten the pack when we were at night stops. I was fed up with kicking his backside over it and decided he would have to learn the hard way.

The night came when Paddy had made a really good Bergen Pie. I noticed the pack was left open, but refrained from warning him. It poured with rain in the night, half-filling Paddy's Bergen. His cigarettes, which had been tipped out of their waterproof bag were a sloppy mess. I watched Paddy trying to salvage some of the cigarettes, knowing there was no hope.

The next day, having smoked all his remaining dry cigarettes, Paddy tried to scrounge some of mine. I refused his pleas for the next twelve days of the operation, only giving him a cigarette when he used the radio and after our evening meal. Paddy's Bergen packing and general care of his kit improved no end from then on.

As far as the enemy was concerned, our operations continued to be mostly uneventful. We were forever going off into nasty places, only to find we were too late, or the information which sent us there was wrong. I began to suspect the Indonesians didn't have any Special Forces. It was all a bluff to make sure my patrol stayed in for fourteen days every time.

Searching for Special Forces is not my idea of a relaxing sport. No matter how good or bad the Special Forces are, they always seem to have plenty of fancy kit, especially in the demolition, mining, trapping and booby trapping departments. Always being interested in staying reasonably alive, I made it my business to find out what the Indonesian Special Forces were likely to have in the way of defence equipment. They had just about everything anyone could think of. Having been trained (and were likely still being trained) by the US Special Forces, they were liable to have just about anything.

In Vietnam the Americans were putting to use just about every diabolical trick known to man and were inventing new gimmicks all the time. We had no doubt the Indonesians would have access to a lot of very difficult-to-beat fancy goods from the American armoury, so we could expect their jungle bases to be protected very effectively.

It was pretty sure they would have Claymore mines available, either the manufactured off-the-shelf job or equally nasty homemade copies. The Claymore is a sort of directional explosive, fair-shares-for-all, ball-bearing spreader, with quite a good range, able to be set off electrically by a sentry or by the intruder touching a trip wire.

I don't know why, but the one thing which worried me most, of which we knew they had plenty, was a very simple little contraption containing one bullet. It could be placed in a small hole in the ground, and, when stepped on, fired its bullet upwards. Very simple to position and virtually impossible to see when covered by one leaf in a sea of leaves! According to some reports, which I saw no reason to disbelieve, the enemy Special Forces were experts in the art of arranging surprises with the aid of local materials. The kind of thing which finishes up with people like me being skewered on a big lump of sharpened bamboo. Or maybe several big lumps. All very well if you know where the base is in the first place, but we never did – except in the case of Kapoet, where there wasn't one! It was always three or four square miles of dense jungle, somewhere nasty to get at and they're in there somewhere. Of course, that few square miles of jungle was always surrounded by a few million more square miles of jungle all looking the same, so where do you start, having found it?

You start by looking for tracks. Circle the suspect area to find any tracks which may head for the base, bearing in mind your information could be a couple of miles wrong and you could wander into the defence perimeter of the place before you find any track. Having found a track, you read it for signs then, depending on what you find, you either watch the track for a day or so – depending on time available – or you start following the track towards the mostly likely direction of the base. But never use the track! It will be the focus of all the diabolical tricks aforementioned and almost certainly be watched.

You push off back the way you have come a couple of hundred yards, then work on your map, what you have seen of the ground and decide where the track is most likely to go. A 'Chinese Parliament' is favourite at this stage – four trained heads being always better than one.

Then you loop around where you think the track will run. Maybe two or three hundred yards, maybe a thousand yards from where you found it. Two things of prime importance must then be borne in mind. First, you may reach the enemy defence perimeter before you find the track again. Second, an enemy patrol will make no noise at all on the track as you come blundering through the bushes. All very sweat-making.

By the time three months had passed I would think we were the only patrol who had not been chased out of the jungle. The others seemed to be getting all the fun. The terrain we covered was typical of the whole area: jungle-covered mountains, with sheer cliffs and deep ravines in places. The going was always hard.

At one LZ right on the border, we made friends with the local tribe, giving some of their members medical help. Their longhouse was close to the LZ and, as it was in a dodgy area, we made an agreement that if all was clear (no enemy around), the children would come out into the LZ and wave to us when they heard the helicopter. This worked well and made life a lot less nerve-racking on two or three occasions. The helicopter pilots appreciated the arrangement too. They always looked relieved when I told them to only approach for landing if the children gave the all clear.

There were other LZs, of course, which were extremely sweat-making. I always felt very vulnerable just before we touched down. Sometimes the chopper hovered just above the ground and we jumped out, in case of mines.

All the jungle people I met on our side of the border were very friendly towards us. They always had been very pro-British and some had fought with distinction against the Japanese during the Second World War, when Borneo had been under Japanese occupation.

Thoughts of the occasional lapses into their old games of headhunting and using their blowpipes to zap poisoned darts at people were pushed from our minds when we got to know them. Nevertheless we were well pleased no Brits had got the jungle tribes' backs up over the years.

Our medics had a field day almost every time we met any jungle folk, with pills for this and stitches for that. Their response to medical treatment had to be seen to be believed. One man I remember in particular had a very nasty five-inch gash in his leg. When we first saw him, on our way in to cross the border, the gash was wide open, about three or four days old and in a terrible mess. The wound was cleaned and stitched. The man was given some antibiotic tablets with verbal instructions on when to take them. When we returned fourteen days later, the man met us with a big grin and no sign of a limp. The gash in his leg had completely healed and looked about six weeks old. Our medicines probably had maximum effect because those people had never had modern treatment before and had not built up antibodies to the medicine.

They were good people to be with, not 'grabbers' like some I know in other parts of the world, and took a gift or any help as an act of friendship – not a god-given right. There was a lot of trust between the jungle tribes and the SAS. Many good friendships had been made in earlier days when time was not so short. We could shake the hand and believe the smiling eyes, ignore the well-used blowpipe in the other hand – and that which probably hung decoratively in the longhouse (often, I've heard say, around a proudly possessed portrait of Her Majesty, Queen Elizabeth II).

Owing to the nature of our work we saw a certain amount of wildlife, some interesting and some dangerous. As far as I know, only the snakes were dangerous. Even they would move away as quickly as possible rather than cause any bother. On one occasion I almost stepped on a cobra. It was behind a log I was stepping over, curled up in a small patch of sunlight. My foot stopped in mid-air about a foot above it then, when I put my foot on the log, it awoke and took off, fast.

Another snake, about seven or eight feet long really made me jump. I was climbing a steep bank out of a shallow stream, pulling myself up over some tree roots, when it came out of the roots right in front of my face. It went over my shoulder, hitting my arm and shoulder as it went. It was dark brown, but I have no idea what type (E-type perhaps – judging by its acceleration!).

There were a few scorpions about. One I saw just in time, and it was as big as my hand, black with a tinge of green, at least eight inches long from nose to tail. I missed a rare sight once when I left the patrol to do a recce on my own. They watched a python attack a large squirrel-type animal. Can't remember the outcome, but they described how the two tumbled from a low tree and rolled along the ground almost at their feet.

Our biggest problem in the wildlife department were pig flies. The little ones in the UK are called Old Maids or Horseflies. The ones in Borneo were all sizes, mostly big. They would creep up on carpet slippers then stamp with red-hot boots. Paddy once lost his temper and slashed at a pig fly with his knife. He only succeeded in cutting our roof nearly in half, much to my disgust.

Another danger in most areas were the pig traps set by local hunters. These consisted of a very sharp bamboo spear attached to a springy sapling. They were triggered by a hidden trip wire. A few of our people had been terribly injured by these devices during the campaign in Malaysia. They could, of course, be set at a height to trap men. The ones I saw in Borneo, however, were definitely set for pigs, being at about knee height or a little higher. Luckily for me I detected the traps before springing them, except one which had killed a pig before we arrived on the scene.

On our way back to Kuching after another uneventful slog, Pete produced from his pack a roll of white material used for weapon cleaning. He induced us all to tie a piece around our left upper arm. Away from the noise of the chopper, he then explained.

Amid curious, suspicious glances we turned up at the Operations

Centre, and, as luck would have it, the squadron commander was the first to crack. 'What the hell are you all wearing white armbands for?'

Quick as a flash Pete ripped off his armband. 'Sorry, sir, we forgot to take them off. That's how we've been getting away with it all this time – we've been going in as umpires!'

Pete's patter was not always good for morale, as in one case I well remember. We had been on a very strenuous and tricky problem, were on our way back towards the border having carried out our task without too much cold sweat and were stopped for the night. Pete was eating his curry and rice in complete silence, which was unusual. Suddenly he turned to me and said, 'Lofty, what the hell are you doing here?' I just stared at him, there was no knowing what might come next. He went on, 'Do you realise, a sergeant in the Ordnance Corps, whose greatest risk is dropping his wallet on his toe, is getting more pay than you are?' Having noticed these strange quirks of army pay a long time ago I said nothing. He rubbed it in. 'The bastard's getting Trade Pay and will be all set for a good job when he gets demobbed. What good is this lot to you? You can't do this in Civvy Street, you know!' I didn't lose any sleep over it, but some people might have, in our position.

Kevin's friends in the Navy offered to take the patrol on a trip in their minesweeper. It seemed a good idea, as we could leave one day and be back the next by changing ships as one took over patrol from the other. Before checking with our people we decided to have a look at the minesweeper and meet some of the crew. Arriving where the ships were berthed we were impressed by their battleship-grey smartness. They were small, as ships go, but seemed quite large to us. With their few guns they looked quite effective. We were welcomed aboard and, as is inevitable, taken to the small wardroom for a drink.

Enjoying my drink I gazed around and noticed slim wooden pegs sticking out of all the walls in a very haphazard pattern. Not wanting to appear ignorant, I contained my curiosity for as long as I could, hoping all would become clear without questions – just before we left, I could wait no longer, and asked about the strange wooden pegs. The Navy types glanced around at the twenty or so pegs around the small room, and one said, 'Oh, we haven't got around to proper repairs yet, they are the bullet holes from our last trip.'

We thought they were joking. A closer inspection of the ship proved they weren't. Soldiers don't think of naval problems much and the fact that – because of magnetic mines – minesweepers have to be made of wood. To us the ships looked bullet-proof in their smart grey paint. It came as a rude awakening to realise these merry matelots were riding around in a bloody sieve. Needless to say, we chickens found strong operational reasons for not taking up the Navy's generous offer of a trip. Our type needs its feet firmly on the ground before taking on spiteful people with guns.

A lack of communication, or mistakes in communication, led to some near scrapes for us on several occasions. One of these was quite lucky, as it happened. The LZs along the border were always dicey places, sometimes downright dangerous. After one particular job we came out at a point on the border some distance from our entry point to an LZ we had not visited before. Having found the LZ we signalled base that we were ready to be lifted out.

There was a long wait while base told us to maintain listening watch on the radio. After about ten minutes they sent a message telling us to move at once to a forward infantry position – a sandbag fortress, about four thousand yards north of the border and about five thousand yards from where we were. The infantry were expecting us, we were to move openly with our hats off, to prevent mistakes (if we met infantry patrols) and be easily recognised as Brits.

Accordingly we moved to a track marked on the map, which turned out to be a once well-trodden path through the jungle, leading almost directly to the fortress. We arrived at the position to find it almost deserted. The infantry officer looked at us in amazement. 'Where the hell have you come from?' were his first words, then, 'Who the hell are you?' I told him we had come from the border. (We were not allowed to mention going over the border in those days.) He asked me if I had seen any of his ambushes. I told him we hadn't. So he asked which way we had come. When I told him he looked shocked. 'No one uses that track,' he said. 'It's a death trap. We ambush it, the Indos ambush it and, in any case, everyone has mined it! I didn't put an ambush on it today because it's a waste of time – nobody uses it any more!'

I told him he was in good company – we hadn't seen any of his opposite numbers' ambushes. 'And while you're thinking about it, you'd better check on your mines, they're not working today either!'

It turned out, as we'd come to half expect from previous experience, the infantry knew nothing about us being in their area. Returning to base we did the usual bitching, which got us nowhere. One of my crew drew some sour looks when he mentioned it was safer going over the border than coming back. 'Perhaps we should ask the Indonesians next time – they seem to know all the safe crossing places!'

Asking why the chopper couldn't make it to our border LZ, we were told it was considered too dangerous, there was too much enemy activity in the area at the time. I got the impression we were to blame for that. One of my voices in the background muttered, 'Another fine mess you got us into!'

Going in on another job, we went by chopper to a forward infantry position which was, I believe, manned by a Guards unit. We arrived just before lunch and were due to be lifted to a border LZ later that day. A very smart, regimental gentleman buttonholed me and said, 'Would you please come with me, sergeant.'

Wondering what was required, I left my crew, who were heading for the cookhouse, and followed the gent to what turned out to be the sergeants' mess, where a huge and beautiful meal was already laid out for me. Never having been one to turn my back on the better things in life – especially when it would offend a most generous host – and having been told the boys would be well looked after, I set to with a will and demolished some very excellent cooking.

Rejoining the patrol when the chopper arrived I got some old-fashioned looks, but nothing much was said. That night when we made camp and went through the usual signals and meal-preparation routine I noticed a few things were different and there was a strange feel to the patrol. Then I noticed a bit of paper stuck in a split stick in front of my poncho. I picked it up; on the other side it said, 'sergeants' mess'. Paddy said, 'If we're not good enough to eat with you out there, we're not good enough to eat with you in here, so cook your own bloody scoff.' It was a send-up which lasted a few days and caused a lot of silent laughter among us.

In operational theatres the SAS always had an all-ranks mess. Everyone from the Brigadier Commanding SAS Group, or the Commanding Officer of 22 SAS, down to the newest young trooper in the regiment took it as normal to sit down together for meals in the same mess – the same table if it happened to be the one vacant. Perhaps it was after that job, when the sergeants' mess send-up was still in the air, we arrived back at Kuching by chopper and landed at Brigade HQ. Well, not quite at Brigade HQ but really just outside, next to a little bar we knew quite well. We had a quick drink to sort of clear our throats, then I went off to ring our people for some wheels to get us and our gear back to SAS base. It took me some time to get through and organise transport. By the time I returned to the little bar the other three were looking suspiciously the worse for the time I had taken. By the time I'd had another drink the wheels arrived and it was then I found they had put all their drinks on my book! The bill was horrific, I could hardly believe so much could be shifted by so few in so short a time, until I saw them struggling to get in the back of a Land Rover like it was Mount Everest!

During that trip to Borneo we were joined on operations by a half-squadron of New Zealand SAS who fitted in with us like very few non-Brit troops can. Their sense of humour and general approach to life were very compatible with ours. Like us, they enjoyed the challenge of operations, pitting their wits and skill against a determined enemy.

Of all the various foreign troops with which I came into contact, the Kiwis were most like our own people. But of course we didn't think of them as foreign. I don't think any of us thought of any Commonwealth troops as foreign, so it was always a good start to getting on with things we had to do together.

15 THE KOEMBA JOB

'D' Squadron's tour of duty in Borneo was nearing its end. We were due to be relieved by 'A' Squadron, and would return to UK for a short break before going on our next job. There was however time for the squadron to have one last try at a job which had already been tried six or seven times by patrols from 'B' and 'D' Squadrons. It was a mental thorn in the mind of our squadron commander. Brigade HQ were warm under the collar about it. We all knew that no job could be termed 'impossible'. In fact it looked dead easy. The main problem was the opposition also knew it looked dead easy, and, as previous patrols had found out, were dead set on making it dead hard!

Everyone on our side of the border wanted to know if the Indonesians were using the River Koemba to build up their troops and supply them. If they were, to what extent? And what could we do about it? Everyone on the other side of the border had seemed, so far, to be very much against the idea.

We didn't even know if the river could be reached along most of its length near the border because of vast swamps in the jungle. The swamps made most of the area operationally impassable. The other bits were well patrolled and more easily watched by the opposition. Reading the reports of previous patrols I got the impression if the swamps didn't get you the Indos would.

For some inexplicable reason the squadron commander got it into his head that my little band of comedians would crack it. (Or was this his final attempt to rid the scene of too much operational hilarity?)

Before briefing me for the job the squadron commander asked me if I felt completely fit and 'up to scratch'. I assured him I had no problems. Then he asked me about Pete, and told me to give Pete the option of dropping out of the patrol for this last job. When I asked the reason, he told me Pete and myself had been continually on operations, mostly difficult ones, for too long – according to 'medical advice' – so it was only fair to give Pete the chance of a rest.

I didn't fancy changing the team, but gave Pete the option anyway. Just as I expected, Pete blew his top. His reply was unprintable, very much to the point – and just what I wanted.

I was given all the gen to study. Then the big question. Did I think we could crack it? I thought we could give it a damn good try, and said as much.

The other patrols who had made the attempt had all been sent to specific areas of the river with boundaries which they could not cross for operational reasons, such as other patrols in adjoining areas or artillery target areas. My first request was for freedom of choice for the route in, and freedom of movement anywhere within reason. This was granted.

I went down to Brigade HQ and my good friend Staff-Sergeant Taylor. We went over those maps and air photos with about everything. All the information we could glean from other patrol reports we related to the maps and photos until I formed an idea of the only possible route which would be worth trying.

There was a spur of higher ground running south from the border towards the river. The spur vanished from both maps and air photos a long way from the river but, following the line of that pointing finger, there was a corresponding bend in the river. It was little enough to go on, but it was all we had. A detailed study of that particular bit of jungle revealed almost nothing. The only 'holes' in the trees on the air photos revealed water. Even Taylor couldn't estimate the depth of water through little holes in the trees. I remembered there were only two holes anyway, in an area of several square miles. There was an Indonesian military set-up, a small garrison base, less than one thousand yards from where I thought I might reach the river. It was on the north bank of the river, the near side, and was probably the base from which their patrols covered the area to the border.

It was obviously on high ground, clear of flooding from the river, so, at a push, I could get close to the base and use their high ground. The main task was to get all the information we could. The secondary task was to disrupt any military river traffic we found. To help with our secondary task I requested a Bren light machine-gun to be carried for added firepower. As I was the biggest patrol member and well able to use a Bren like a rifle, I thought it was a good idea. The squadron commander, however, thought the extra weight and more cumbersome weapon just might tip the balance the wrong way and interfere with our primary task. The Bren was not granted. He did suggest I take an AR16 (Armalite) but at that time we only had two in the unit. One was being carried on operations already and the other – you would never guess – had been taken to Labuan, a safe base area hundreds of miles from Kuching, by a gentleman who, in that situation, didn't need a weapon at all! Having a fancy gun to parade around would do his ego a power of good, but very little to help the unit effort in the field.

I counted it a small price to pay for peace and quiet, so, although a few alternatives were available, I opted to take my tried and trusted SLR. At least it had the punch we wanted, if not the rate of fire.

The patrol worked out on the jungle training areas, firing and checking our weapons, as well as all our contact drills and operational procedures. By the time we set forth we were well pleased with every detail. Confident in ourselves, in each other and in our ability to do the job. We knew we were going to reach that river – come hell or high water!

Our departure was marred by something we could do nothing about. Our weapons had been checked, cleaned and oiled to perfection, as our lives were likely to depend on them. They were then handed in to the squadron armoury for the night and returned with the vehicle which picked us up to take us to the airfield in the morning. It was several miles from Squadron HQ to where we were billeted, then a few more miles away from SHQ to the airfield. When the vehicle arrived we piled in and immediately grabbed our beloved rifles. There was a yell at once from Pete: 'Where's my rifle? – I don't want this bloody thing!'

Somehow the wrong rifle had arrived and Pete didn't want to know it. There was no time to go back for the right weapon so we had to take it. I told Pete it must be okay. We would try to test-fire it at Lundu – where we were flying by fixed wing aircraft, a Twin Pioneer. In the event there was no time, as we went straight from the 'Twin Pin' to a chopper which couldn't wait because of a tight schedule. Arriving at a very dicey LZ right on the border in a hot area, there was no chance of test-firing the weapon. Pete went into the jungle a very unhappy man that day.

That particular LZ was one of the more sweat-making ones. It was well positioned and had been well cleared, but the only place the chopper could be put down was on a sturdy, well-built log platform, about eight metres square. Due to various reasons, such as a considerable amount of enemy activity in the area and the likelihood of having to use an LZ to bring in a back-up force if we got into trouble, I had been ordered to check the platform and the immediate area for mines.

The chopper pilots always liked to get in and out of these dodgy LZs as quickly as possible. And, for our part, we wanted to be clear of the LZ and into the cover of jungle even quicker. In jungle warfare one feels terribly exposed in the middle of an area almost half the size of a football pitch. But, orders are orders, so I had to go through the routine of dangling from the chopper, then dropping ten feet to land in a heap about twenty yards from the platform. The chopper then backed off and went up to nearly one hundred feet so the pilot could see me while I went through the usual checks, looking for anything nasty or suspicious. The absolute vulnerability of our situation urged me to rush, but the necessity to make sure it was safe prevented any corners being cut. Having found no problems I signalled to the chopper to land and within about twenty seconds we were into the sticks and the sound of the chopper was fading into the distance. Blessed sweaty relief! The relief was, however, somewhat tempered by the cloud of doubt which hung over Pete's rifle or, perhaps I should say, the lack of Pete's rifle.

None of us was too happy with the situation. We worked together like a well-oiled machine, there was no room for a suspect part. Pete had stripped the weapon on the aircraft, checked every bit. Tested everything that could be tested – short of firing it. But he distrusted the damn thing, losing no chance of letting the rest of us know his feelings.

The first night in the jungle we all checked the suspect weapon. I for one pronounced it perfect, advising Pete to ignore the possibilities, we had enough problems already. Pete said he felt better about it, but I knew him too well to believe him. He knew as well as I did that perhaps the rifle might not fire as it should – it might not even fire at all – or, even worse, it could fire when it shouldn't.

We knew the area was extremely difficult for navigation: very dense jungle with undulating ground and lumpy little hills which didn't show on the map but were very misleading on the ground. We wanted to make sure we were east of the point where the invisible spur reached the river, so that once we hit the swamps we could keep moving west until, hopefully, we found a way through.

Information from another patrol's report (Old Joe's) told us where the best route lay for most of the distance, so I kept well away from it. We might need it on the way back. There are many problems with jungle movement under operational conditions. One is tracks. No matter how good you are, you are bound to leave tracks in most places. So, on the first day or two we moved on a steady bearing towards a point west of where we were really heading. After that we made for a point east of our real heading. After that we hit swamps and our choices finished. We were limited to where was possible.

All the way from the border we often found tracks of enemy patrols. I realised the whole area was rather heavily patrolled. This, of course, was only to be expected, bearing in mind the experiences mentioned in other SAS patrol reports.

Several times we made small detours away from sounds I couldn't positively identify. There was no more than the usual amount of wildlife, but occasionally monkeys or birds (probably) rustled branches or made some kind of noise without any other sound to identify the source. With all those tracks around I took no chances. On the other hand, a very silent jungle is also a danger signal at times. The question then arises, 'Why is everything keeping so quiet?' and 'What do they know that I don't?'

On the second day, I heard a slight sound up ahead. Leaving the patrol I went forward and found an Indonesian unit of platoon strength right in our path. We pulled back a little, made a detour to the east of them and pushed on. The enemy platoon were stationary. Some were cutting small branches, making the noise which gave them away. So we had no idea which way they were heading or what they were doing. It was too early in the day for them to be making a night-stop camp. There was no track anywhere we could see which they might be ambushing. As we didn't find their tracks we knew they must have come from west of our route, so it was possible, or probable, they would cross our tracks when they resumed moving. Not good news. We did what we could with our track problem. Kevin keeping an extra sharp look and listen to the rear.

The edge of the swamp was just a matter of splashing in and out of areas of shallow water on to areas of low firm ground. Further in, the water became gradually deeper, the ground above water level became less firm, then just mud. A jungle swamp is not a nice place. Under the black water, tree roots tangle and jam unwary feet. Deep holes are not usually common, but they exist. The jungle is still there above and below the water, in all its tangled glory, but now decorated with hanging moss and drapes of jungle rubbish up to the high-water mark; in this case several feet (up to nine or ten feet) above the present water level. Fallen trees and branches still impeded progress, but were no longer visible.

Sometimes a thick green scum covered the surface of the swamp over large areas. This particular swamp had an extra hazard: huge leaves, mostly four or five feet long and two or three feet wide, which had fallen from a type of tree which was apparently very common in that area. The leaves crackled loudly when stepped on and were even noisy when moved to one side on the water.

By the time I had been waist-deep for about half an hour at a stretch, trying in vain to move silently southward, I had this recurring vision of a grinning enemy watching me approach noisily to the muzzle of his heavy machine-gun. The water crept above my waist and I stopped. The decision, silently made a long way back, was that no way was I going to fight a war more than waist deep in that lot.

Turning on to a westerly bearing we moved until we found dry ground. There we took a break, having spent about ninety minutes in the water. Leaving Paddy and Kevin to look after the packs, Pete and I went south again, looking for a way through.

We were probing the east side of where the spur should be – if it still existed – and would continue pushing south towards the river as far as we could on each probe, moving west after each failure. At one point I came to a fallen tree, about a foot thick, lying in the water. It was just under the surface, as I reached it and prepared to climb over, the water swirled on the other side and I glimpsed a large patch of light-brown scales going down under the tree towards my legs. I was nearly airborne by the time I realised it was only a turtle or tortoise.

After about an hour, we stopped in waist-deep water while I checked the map, compass and memory for anything which tallied with the terrain we had crossed.

Pete was probably as knackered as I was; it was bloody hard work. I glanced back at him, he was about twenty yards away, right where he should be. Seeing my glance he whispered something. I couldn't hear the words so signalled him closer. Closing up to about ten yards, he whispered, 'I hope we're the goodies!'

In a daze, I stared at him. 'What?' He repeated the same sentence. Losing patience fast I said, 'What do you mean?'

Pete closed up a couple more yards, his face and eyes serious, 'Well,' he said, 'if we're the baddies, we'll just as likely finish up with our hats floating on top of this lot!' Then he let the grin slip, and I had difficulty in keeping the laughter silent. After that I felt more relaxed, and inwardly blessed that comedian for his sense of humour.

For two or three days we pushed and probed, then we cracked it. We had been so close on occasions, we had heard the heavy diesel engines on the river, often within one hundred yards of us. When we got through, we were all moving together as a patrol and I could hardly believe my eyes. There were the muddy swirling waters of the elusive Koemba. Part of the shock was seeing it about seventy yards away, down a steep slope through a rubber plantation. I checked the area, then brought the patrol forward so they could see what had stopped us and deploy to cover me while I checked all around. The rubber plantation was not in use, although cups still hung on the trees. Tappers' tracks were old and overgrown. There was a track along the top of the bank from where I first saw the river but that too was disused.

A few hundred yards earlier we had crossed two, fairly fresh, cut tracks. They were about one hundred yards apart and, on taking bearings, I found they were almost exactly parallel to each other with one being quite wide, about three or four yards. Neither showed signs of use, but as the bearing was a little south of west I guessed they went almost straight towards the enemy base on the river. There are various reasons for that type of cut track in jungle warfare. The two main reasons which might affect us were deployment tracks in the one instance or field of fire in the other.

In the case of the wider track I estimated it served a dual purpose, allowing troops to be deployed quickly and, due to the gently undulating ground in that area, it would be ideal for fire groups to be stationed at pre-determined points to watch for, and fire upon, anyone crossing the track. We had our own methods of dealing with these problems but knew they would have to be kept in mind on our way back.

As I moved forward to check down to the river, the whole plan became clear. I signalled the patrol to follow as soon as I knew it was safe. They came down through the rubber taking care not to leave tracks. The place was made to measure. There was an old ditch about three feet deep and ten feet wide, just short of the river bank. A few bushes hid the ditch from all sides. We moved in.

In front of us now was a good view of the river as we were right on a sharp bend. The bend would slow and turn river traffic so we could get a good view. The ambush possibilities were beautiful. Our only problem was our tracks. We thought back over our last bit of dry land after the swamp and decided it was worth taking the chance. To go back and look would only create more problems so we decided to stay put.

The river was roughly forty yards wide with the nearer side being a deep channel against a high vertical bank. It flowed west, left to right as we looked at it. Our side of the bend was about thirty yards to our left but, due to the width, we had a good view up stream for one hundred yards. The far bank had no large trees and, being much lower than our side, we could see it was obviously often swept by floodwater as bare rock and large pools showed through the thin cover of bushes and small trees.

At first, it puzzled me why the rubber plantation didn't show on the air photos. Then, when I looked around, I could understand it. The rubber trees were planted haphazardly, not in rows. There were odd jungle trees growing among them down at the river edge and along the lower slopes. The higher ground meant the rubber trees grew up to the height of the jungle near the edge of the plantation, giving an overall flattish canopy right across.

Assessing our position, we decided our weak spot was the high ground from which we had first seen the river. But the only time we would be exposed to view from that side was when in our actual ambush firing positions. On normal watch there was no problem.

The whole situation tickled our sense of humour. Morale was sky high. What kind of fool would hide under a few bushes in a rubber plantation, within a few hundred yards of an enemy base, when there was a whole damn jungle everywhere else? It looked ridiculous as a hiding place. Enemy patrols would concentrate on the jungle, they wouldn't give our bushes a second glance, even if they came anywhere near. Night came, we cooked a good curry and slept.

Top exponents of jungle warfare and young lads wanting to learn the art – each group trying to impress the other – should read no further. Very likely on real pain of death. From apoplexy in the one case and from trying to do the wrong thing in the wrong place in the other.

To the good, tough, resolute fighting men of the British Army who had suffered ambush positions of complete silence, no smoking, no hot drinks, no cooked food (no farting?) for days on end, sometimes a week or two, maybe more, the hard truth (never told before) which follows may be too much to bear or believe. I could cloud the whole scene in bullshit, without mentioning realities, but so far I have stuck faithfully to the truth of my memory and see no reason to change.

We all knew the rules laid down by the regiment, concerning ambushes, observation posts, movement in enemy-held areas and a whole bible of operational situations. I make no excuses. I had my reasons. I know I have been unorthodox, but on purpose, with a logical reason.

Our position would be good until the shit hit the fan. We had passed the 'cut off' tracks, freshly cut in nice straight lines. We had seen sign of enemy patrols and almost bumped one. We were not far from an enemy base and knew we had to go nearer to it when we bugged out – if we were to make a fast escape from the potential trap we were in.

I wanted fitness and morale at its peak for the bug-out, which we all knew would come. I wanted a cock-a-hoop patrol which would take on the world with a grin. So I got one.

We smoked. Even Pete, the non-smoker, had the occasional cigarette. We had a brew of coffee or tea about every two hours. We had a good meal of curry and rice every evening. Relaxation, as far as is possible under those conditions, was the order of the day. We got most of the information required in the first twenty-four hours. Although we had to stay put as long as possible to gain as much information as possible, we were not too bothered about picking a fight to finish the job any time anyone wanted to try us. We felt very secure in our own confidence.

The first thing we saw on the river nearly finished us. Kevin was fishing for water. He had his water bottle on a cord dangling it in the river, which was about seven or eight feet below the sheer bank in front of us, when a canoe carrying two men came silently upstream from the west. They came just past the position then turned to face us. I watched them closely. They could have been locals, but I didn't think so.

One was working his paddle hard and staring at the bank just to the left of me. The water-bottle cord was away to the right, thank God. The other man was staring into the muddy water and working with a stick, which turned out to have a hook on the end when he pulled up a fish trap. Situation explained.

I continued to watch them closely as they moved upstream and lifted other traps. There was no sign they had seen anything of us, so we let them go. Being a suspicious lot, the patrol had all been looking the other way, in case the fishermen were being used as a distraction.

Several diesel launches had passed us during the second day. Heavily laden going upstream, empty on their way back. Then an open boat with an outboard motor appeared. There were three men in it, all wearing blue shorts. They looked too big, too military in appearance to be locals, but I wasn't sure.

They beached the boat about twenty-five yards upstream from us, then went to work on baling out the boat and refuelling the engine. The patrol was in an all-round defence posture. I watched the men closely. I can't remember seeing any weapons, but there could have been some among the kit in the boat.

I thought of taking them prisoner. They could probably give our people all the information they wanted if we could get them back over the border. Three or four things were against taking them prisoner. First, I wasn't sure they were military. Second, it would be difficult enough to reach the border even without a prisoner. Thirdly, they might have had no knowledge of use to us, whether or not they were military. Fourth, I had explicit instructions not to initiate any type of offensive action until my primary role was completed and I received authority from base. So they finished their work and went on their way, never knowing how lucky they were.

The days passed. We informed base of everything which moved on the river. On one occasion a raft came down the river with three or four men on it, all singing. One was doing the lead singer's job quite well, he had a good voice. I glanced back at the patrol. They all had their arms waving in the air like they were conducting an orchestra. Morale was high.

On the third or fourth day morale was still at its usual high level, though we had probably lost a few pints of sweat from one nerve-wracking encounter: a hunting party with dogs made a sweep of the area on the other side of the river. At a guess, there were five or six men, and at least two of them were carrying shotguns. They were not in uniform as such, and although one was wearing army slacks the others I saw were in shorts. They were all stripped to the waist. The dogs were running free – which was a bit worrying. One was a small labrador, another was a doberman, or something similar, and there were one or two others which sounded small and yappy. The men were definitely not jungle folk; probably off-duty personnel from the nearby base. We hoped they were not intending to sweep our side of the river for a week or so. The jungle is full of surprises, but a hunting party complete with bloody dogs . . . That's all you need when trying to do your job in a dodgy patch of bushes among the opposition. We were prepared for them being used as a distraction or recce group so we stayed on full alert for the rest of the day and then from dawn to midday on the next day. Anyway, no problem, we didn't see them again.

The time came when I calculated we had done enough. Time was running out, so I sent a signal requesting permission to carry out our secondary task. Being of an extremely distrusting nature, I didn't trust the opposition not to have the necessary radio equipment to get a fix on our position when we used our radio to send messages. As a result, all our messages were as short as possible, with Paddy and I spending a lot of time finding the shortest possible way to code each message.

It was 1965 and James Bond films were at their height. Checking for a short way through the codes, we found the shortest. The message read 'Request "00" Licence'. Everyone at base would be expecting the request. I thought it was unmistakable and everyone would realise that 00 – as in 007 – would mean 'licence to kill'. Everyone at base understood exactly what was required – except the clown who was in charge of the Operations Centre at the time. With everyone else in the place telling him it was obvious, he sent back a signal saying, 'Message not understood'. The difference in time sending the next shortest alternative coded message was about twenty seconds instead of the original four or five seconds.

There is no way around stupidity in high places when you're out on a limb. I only wished the squadron commander had been in the Ops Centre at the time, but he wasn't. We sent the request the hard way, and hoped (as we always did when working to base from close to enemy positions) that no enemy signals type was getting his pop music interfered with by our morse code. Having penetrated the thick skull with a long hammering message, we received permission to proceed with stage two of the operation.

Now it was a matter of finding the right target at the right time. I wanted the biggest possible military boat moving upstream in the late afternoon. The rest of the patrol wanted any floating object as soon as possible and to get the hell out of there.

We agreed to a deadline of one hour before dusk. If we had not found a target by then we would wait until the next day.

Our rucksacks were packed and ready to go, as they had been at all times during daylight. Each of us had one day's rations in our pack plus our normal escape rations on our belts. The other rations, cookers and rubbish were in plastic bags, each with a heavy rock to sink them in the river after the ambush. We only wanted to be carrying the bare essentials when we bugged out. Paddy and Kevin would dispose of the unwanted bags immediately after the ambush, before they went up to the edge of the jungle.

Our big problem was not the ambush, but the getting clear away after it. The enemy base just down the river would be sure to hear the attack and could well be in a position to send troops along the deployment tracks to the north of us, thereby cutting off our escape. If they reacted quickly they could reach the crucial point before us, as they had less distance to go and good tracks to move on. There was not a lot of room between the swamp and the open rubber plantation, both of which we must avoid. If the enemy could get troops to the north end of that strip of jungle we would have big problems.

Not knowing how far north the rubber trees went was another problem. There had been no sense in doing a recce of the area, so the only part we knew was our route in, which was close to the swamp. So that was out, as our tracks may have been found there. This meant we had to go further west on the way out, nearer to the edge of the rubber plantation, also nearer to the enemy base, thereby shortening their distance to the likely point of contact and lengthening ours.

I estimated we could get past the deployment tracks in about half an hour if we moved fast. Once out of the bottleneck our chances would be good – if we could keep moving fast.

I also estimated the enemy would deploy without their packs, being so close to their base, so would have two chances late in the day: either to return to base before dark, or to spend the night cold and miserable in the jungle. Either way we would have the advantage the next morning – if we had cleared that area. The enemy patrols further north might, or might not, still be there. Luck would likely be the major factor in their case, so there was little we could do about them. There was also the chance we may get into problems and be glad of darkness to cover our escape. The jungle is not impassable at night in that situation.

A straight beeline for our extraction LZ (the same one we entered by) would not be wise with those patrols to the north, so I planned to put an easterly 'crank' in our route, partly to avoid interception further north (which could be organised by enemy troops following our tracks and sending the position and bearing to those ahead of us) and partly to take advantage of the good going, mentioned in Joe's patrol report. The patrol had haggled through all the pros and cons. We were all satisfied we had taken care of all foreseeable possibilities (plus a few others). We all knew what to do once the ambush was over.

So we waited. Noon passed and the optimum time was approaching. Two or three launches passed. They were too small or going the wrong way. Those going downstream were obviously going faster and would quickly be carried out of sight. The downstream traffic came straight towards us, then turned and soon vanished to our right. The upstream traffic came slowly from our right, into the bend, then slowly away from us, to disappear around another bend about one hundred yards away.

The actual ambush, if we picked the right target, would be unlikely to give us any problems. Our firing positions were planned accordingly.

We would move out of the bushes, along the ditch to the left, where it was much shallower, and fire from positions lying against the bank nearest the river. At that stage we would be completely exposed from the back and left flank and as we moved to the firing positions we would be in full view from the river. I would move out first and go further out than the others, to a position on the rear bank of the ditch, and remain standing so as to be able to see all around the area, to give early warning of any other problems arising, such as an escort boat following the target or enemy appearing above us at the edge of the jungle.

It was also my job to take out any possible source of retaliation from the target. We hoped to judge things so the target boat would be just starting its turn on the bend and at a range of twenty to thirty yards.

Our plan was to destroy a boat. We were not too interested in inflicting casualties. There had to be casualties to prevent return fire but the main fire-power would be directed at the stern of the boat, just above the waterline, hoping to destroy the propeller drive and hole the bottom of the boat and also destroy the engine.

We knew our rifle fire would penetrate over two feet of timber at close range. This was very handy to know in jungle warfare as people tend to take cover behind trees. So, if the tree was only about two foot thick we would just fire at the tree. In this case, with a boat, as we were roughly twelve feet above the water level and at very close range, there would be no doubt about penetrating the hull. A concentration of rounds hitting at water level would do a considerable amount of damage by the time it had gone through and out the bottom of the vessel.

The other patrol members would concentrate their fire on the centre of the stern. Firing would commence as each reached his position and found the target, All shots would be well aimed, there was no rush. We would each fire twenty rounds into the target. In the unlikely event of unexpected opposition after the main event we would keep firing until it was taken out.

Mid-afternoon came, and with it the rain. At first a gentle pattering, but distant thunder promised a heavier dose was on the way. My thoughts turned to the swamps. Heavy rain would spread the swamp area considerably, very likely making our exit much more difficult by narrowing that bottleneck. That's if we were still there next morning.

The roar of the rain muffled the sound of the launch as it approached upstream. I was on almost continual watch, as the decision to attack had to be mine. Peering through the intervening greenery and the rain I saw a new sight on the river, a gleaming white launch with glinting brass-work. This was no ordinary run-of-the-mill river craft. It had to be something special. I could feel the tension of the patrol behind me, waiting for the signal. The launch was nearly upon us and my right hand was held out ready to turn the thumb up or down. Then I saw the woman. She was standing on the bridge of the launch, between two men in uniform. One in Navy whites, the other an army officer.

The bridge of the boat was all glass, except the lower three feet, and it must have had a glass roof as the people were not getting wet. The height of the bridge put them very slightly higher than my position, which was about twelve feet above the water level. Had they looked my way we would have been eye to eye at a distance of about ten yards. The target was perfect, just what we wanted. The launch was a good forty-five foot long, a nice boat, very naval or military in appearance, its Indonesian flag and various pennants flying proudly.

Neither I nor any man with whom I served in the British Army had ever intentionally made war upon women and children. I saw no reason to change. The thumb went down. I watched the people on the boat flash past, thirty or forty feet in front of me. They were laughing, enjoying a joke, the woman's black hair contrasting with her white dress, oblivious of sudden death just a few yards away.

As the launch disappeared from view I heard someone mutter, 'What's he bloody waiting for? The frigging *Ark Royal*?'

The patrol had only been able to see the back of the launch as it turned and went away from us. When I told them there was a woman on the bridge, so very likely others as well as children on board, they were in full agreement with the decision not to attack.

Again we waited. The tempo of the rain increased. Thunder rolled and crashed all around. Lightning flickered and flashed in the dim light of the storm. Conditions were perfect, perhaps too perfect. I wondered if river traffic might call a halt until the storm passed.

I can't remember whether I first heard or saw the next launch. We were almost at the deadline for that day. I was beginning to think of tomorrow's luck when suddenly it was there. It was coming steadily upstream, a big boat, at least forty feet long. Canvas drapes were rolled down at the sides, because of the rain, so I couldn't see the cargo. As it passed me I saw men in jungle green sitting the same way as you often see troops in the back of army lorries. The thumb went up and the patrol hit their firing positions.

The soldier facing us looked up, open-mouthed and wide-eyed just before my first shot, which was all his. My next shot was placed left of him as, although I could not see what was there due to the darkness under the canopy, it was likely there was another man sitting next to him, also facing in our direction. The other man in view, with his back to us, had not visibly moved before the third shot took him out. Once the obvious threat was removed I placed five shots into the cargo compartment, hopefully to take out any others who may have been along the side seating.

By this time the boat was almost stern on to us, having almost turned on the bend in the river. There was no sign of any retaliation, no sign of any movement on the boat.

The patrol was firing steadily, hitting the flat stern of the boat, centre of the waterline. Pete was having trouble with that bloody rifle. I watched him go through the drills, trying to clear the problem, then I put another five shots into the boat to keep them busy, and turned to have a good look along the high ground behind us. Nothing showed, so I again turned my attention to the boat. Pete was still having problems, having to cock the rifle after each shot.

The boat slowed and lost way, smoke began to billow from the darkness under the canvas roof. I spaced six rounds across the middle of the target area, then changed the magazine. Only Pete was firing so I yelled 'Stop!' and that was it, we were on our way. Paddy and Kevin grabbed their packs and ran to the high ground, while we covered them. Then it was Pete and I who ran. But halfway up the bank I remembered my brand-new water bag – a type very hard to come by – so, begrudging its use by the opposition, I ran back to get it, followed by Pete who thought I had seen something up ahead and was bugging out in the wrong direction. He looked fit to bust as he watched me empty the water bag and stuff it in my shirt.

Then we were away. My last sight of the boat saw that it was dead in the water, beginning to turn as it drifted back, almost obscured by thick smoke. Something man-size toppled like a log into the water from amidships. The boat had a list to starboard and the orange glow of fire was visible under the shadowy roof.

About one hundred yards into the jungle we stopped to check everything was okay. Everything was – except Pete's rifle. He was savage as hell. We were all savage as hell come to that. It was bad enough having our fire-power depleted in the river ambush, but it could be lethal for us if we had a head-on contact with the enemy while moving fast. Especially for me. I relied completely and with confidence on Pete to get me out of trouble if we hit anything. His fast reactions and ability with a weapon were crucial.

We agreed on adjustments to our contact drills which ensured we could do our best, come the crunch, and headed out fast. With any luck the uproar of the storm had covered the sound of the ambush, but it wasn't something we wanted to bet our necks on. The chances the enemy would be waiting for us were about fifty-fifty (much better than the ten to one odds against us without the storm).

The advantages and disadvantages of slow or fast movement in different types of jungle have to be weighed against the requirement of the moment, the likely enemy situation and the type of jungle encountered. A considerable amount of know-how and experience is required to get the right balance in the right places and at the right time. An enemy lying in wait has everything going for him – except when! He's ready to blow you in half but never knows when you will appear, if ever.

Trying to move up silently on someone who is switched on and expecting you is one thing to avoid in jungle. It is almost impossible. Smashing through the sticks like a rampaging bull elephant gives ample warning of your approach and ensures a hot reception.

We were now going to move as fast as possible, because we had to, but that didn't mean we were throwing caution to the winds, or making things easy for the opposition. To do it right, every member of the patrol has to be very good at this job to give the leading scout that hairbreadth of time afforded by surprise.

The leading scout has the job of finding the way through, as quietly as possible, while still navigating to maintain overall movement in the right direction. At the same time he has to be ready to react like lightning. The only way I could do it was to fix in my mind, 'The bastards are there – go get 'em!' To think that they will appear as surely as the hidden targets on the jungle range – and go for them. Confidence in your own abilities, training, experience and extremely fast reactions under pressure can help a lot. But you still know that if Lady Luck is not smiling upon you – the rest is rubbish.

Every patrol member must be just as switched on as the leading scout. Their extra problems are that their friends are probably between them and the enemy. They must be ready to bring instant fire to bear in the right places while being careful not to hit their own men.

They must keep their eyes open to left and right for any sign of problems, and make sure the man behind is still where he should be. They must keep in sight of the man in front but be far enough back not to be caught in the same burst of automatic fire, mine, or booby trap. In a contact with the enemy they must react instantly, guided by the actions of the man in front of them, or what happens to him. Complete, unwavering trust in each other is essential to any patrol on operations.

Lady Luck had smiled upon us in the ambush by providing us with a thunderstorm at the right moment. She now continued to smile to the accompaniment of pouring rain which would cover the sounds of our movement. Nevertheless, fickle lady as she is, the rain could stop and a platoon of enemy could be waiting right in our path. She continued to smile.

The patrol had almost reached the point of greatest danger, the deployment track at the north end of the bottleneck, where our route lay nearest to the enemy base. I was moving swiftly up a short sharp slope when a movement ahead stopped me.

It was a snake, on the flat ground at the top of the slope. We saw each other at the same time, meeting head on, both in a hurry. The rifle was lined up on it before I knew what it was that moved. It stopped at the same time as I did, and reared up.

I had never seen one before, but I knew only too well I was facing a king cobra. At a range of about six or seven feet, it reared up to a height of at least four or five feet. As it was on slightly higher ground it looked enormous. We were eyeball to eyeball. The rifle had followed it up and was aimed at the centre of the hood, which swayed slightly but was almost still. I remained perfectly still, with my brain in overdrive. I dare not shoot except as a last resort. A shot would not only give us away to the enemy, but would put the patrol into its head-on contact drill. In other words, create a small war on a snake, with a lot of ammunition being expended and possibly quite a delay to further progress in the right direction.

How far can the damn things reach when they strike? Is there any warning they are about to do so? How I wished I'd learnt more about snakes. We stood there, both poised to strike, not knowing each other's abilities and, hopefully, neither wanting to start anything.

It seemed an age that we froze like that, but was probably less than ten seconds, then, as quickly as it had risen, it dropped down and rushed away behind a log. Beautiful, sensible snake. I looked back at Pete, signalled 'snake', pointed to the log and moved on past it. The snake had vanished.

With that danger gone I now had time for another worry. What the hell does a king cobra flee from? It had certainly been travelling fast when we met. Another split-second and we would have collided, but why? The only thing I could think of was man; and the only men in front of me for a long, long way ahead were not the ones I wanted to meet either.

It looked like an excellent idea to wriggle away behind a log – as per snake – and await better times. In spite of our situation I had to grin to myself at the thought of what the comedians would have made of that excellent idea.

We rushed on, reached and dashed across the deployment tracks, then began to breathe easier. No sign of the opposition. Our method of crossing cut-off tracks and the like was simple. If we crossed singly it would give time for any enemy watching the track to fire at the second or third man, having seen the first. So we all lined up along the side of the track, each making sure he had a good hole in the jungle on the other side in which to disappear quickly. Then, at a given signal, everyone rushed across together before anyone knew you were there. This can sometimes draw ineffective fire from inexperienced troops. The worst danger is that you have been seen, so the enemy knows where you are. To help with that problem it is best to make sure the track is still clear after you have crossed; also that everyone of the crossing patrol looks to their right or left as they cross.

The only remaining problem is the usual one of tracks. It is almost certain the enemy follow-up force which came after us on this particular occasion picked up our tracks when they checked the cut-off tracks.

Easy for them as it was our only way out – and they knew it.

The rain, which had given us such good cover, both for the ambush and for the mad dash through the bottleneck, had slackened off by the time we took a five-minute break.

I checked back along the patrol. Reaching Kevin, he asked me what the sudden stop had been for. I told him. For some reason Paddy hadn't mentioned it at the time. Kevin looked past me at Paddy and whispered, 'You Irish bastard – I could have been bloody scoffed!' Kevin was not at all fond of snakes, unless they were curried.

Pushing on fast until almost last light, we were well pleased with our progress by the time we looped and lay among some fallen trees, ambushing our back track. Then we heard the mortar firing from the enemy base. We listened with bated breath for the shells to land. They were a long way from us. Someone said, 'Thank Christ for that, they think we've gone further east.'

I hated to spoil that peace of mind but had to say it. 'Why do you think they're not shelling this area?'

There was a silence, then someone muttered, 'Shit!'

The signal we sent that night got me a sharp dressing down when we arrived back at base in Kuching. Not because of its content, but because of its classification. Paddy had warned me it should have been sent as a top-priority signal, but we had done our job, got clean away (so far) and had no immediate problems which were not obvious to the people in base, who were well aware we were in a tricky situation once the ambush was sprung.

Sparing a thought for all the other patrols who could be in worse situations and possibly needing urgent action to get them out, and feeling in no more deep shit than we had on several previous occasions, it's a wonder I didn't classify the message as routine. So we took the medium line and made it priority, not realising the reports of all contacts with enemy forces to be sent top priority were orders from way above the SAS. I had also thought the lack of top priority for the signal would convey its own message: we were in good shape, no problems, no sweat.

After dark we cooked a good meal, then settled down for the night. We didn't all take off our boots and belts. It was a free choice, I took off mine and crawled thankfully into my sleeping bag.

Before first light in the morning we were ready to go, waiting for the first glimmer of light to give us enough to move. We had each drunk a cup of coffee, were ready for the day and anything it might bring.

Paddy was the only one with any weight on his back, as he was carrying the radio and would not let anyone else share the load. 'None of you bums could use it properly anyway,' was always his final argument.

Once we got under way again I kept check on Paddy as he had had some sort of bug before the operation and wasn't as fit as he might have been.

We put in a distance to the east-north-east soon after starting the day's march, which quickly put us away from our original bearing and into Joe's 'good primary jungle', where we were able to move faster and quieter. Once or twice animals or birds gave me a start and stopped us dead in our tracks, but of the opposition there was no sign. Their tracks showed in a couple of places, but the freshest ones were days old.

Our next problem was finding and checking the LZ, a task which cannot be done in a hurry without risk of springing an ambush.

Base told us by radio there were no friendly forces in the area. I remember hoping they had it right this time, as we were travelling fast on a hair-trigger. I had already come within an ace of shooting a small deer, a bird and – would you believe – a falling leaf. At less than one thousand yards from the LZ we stopped to give base our ETA (estimated time of arrival). I noticed Paddy was getting to look rough. He had almost stopped sweating, a bad sign under those conditions. He said he kept feeling dizzy when Kevin accused him of 'walking like a drunk!'

I had to rethink the angles. We may yet have a battle on our hands. A sick man was something we could very well do without.

There was another problem: our particular LZ was exceptionally difficult to find. Some people might think that the 'magic' SAS were slipping a bit if it couldn't easily find an LZ. In fact the LZs in Borneo were purposely difficult to find, usually having no track or stream near them and well away from likely routes. If the opposition found them it would be sheer chance.

There could be up to fifteen or twenty LZs in an area the size of Dartmoor. Each LZ would average roughly eighty yards by about one hundred yards. It was like looking for a group of six tennis courts on a dark night, somewhere on Dartmoor. Due to the dense jungle it was possible to pass within thirty yards of an LZ and not see it. Add to this, the fact that when giant trees fell they often took other trees with them, thereby creating a small clearing, letting in the sunlight which, from on the ground, could easily be mistaken for an LZ. On the other hand an LZ could just as easily be mistaken for a fallen tree.

There was absolutely nothing between the river and the border which showed on the maps, except the spur, which was almost impossible to find on the ground due to the lumpy nature of the whole area. We had nothing on which to check our position since leaving the river. As it was likely that there would be an enemy follow-up party, possibly (by my estimation) within an hour's march behind us, there was no time for errors. They could have been a matter of minutes behind us, and with that thought in mind Kevin was way back on our back track, watching and listening, ready to slow them down a bit. I made a quick decision. A few minutes now could be well worth it later. We sent a signal requesting a helicopter to come and hover over the LZ for fifteen seconds so we could get a definite fix on our approach. The request was granted almost at once and we were quickly on our way.

When I estimated we were about five hundred yards from the LZ a chopper came over, but instead of hovering over the LZ it circled around as if searching for us. I saw the chopper through a gap in the trees and could see the crewman leaning well out, looking down. I knew they were looking for us, so I took out the homing device and switched it on. The chopper came straight over us and the winch cable was lowered. It was too short to reach us, so went back up then came down again with an extension on the end. This time it reached, so I told Pete to go first as he had the useless rifle. With the extension on, the winch couldn't be fully wound in, so Pete was dragged up through the trees and away, swinging and spinning under the chopper. His eyes were like chapel hat pegs.

The chopper went off a short distance then returned; it was Paddy's turn. But the winchman leaned out and threw down an empty beer can with a note tied to it. The note said, 'Move one hundred yards east.' This we did (and in so doing crossed our outward bound tracks on the way). The chopper had found a place it could get low enough to winch us up properly.

Paddy went first, then Kevin. At that point I felt very much alone. The noise of the chopper was deafening, and the rotors were causing a maelstrom of leaves, twigs and other jungle debris. The senses on which I had so recently relied were numbed and useless. Almost blinded by flying debris but trying to look in all directions at once, I crouched on one knee, the rifle butt in my shoulder, the safety catch off. Everything was moving, low branches and undergrowth alike. I cursed myself for not foreseeing this situation and fixing some signal – like switching off the homing device – for the chopper to get to hell out of there if the opposition showed up. If they did arrive there would be no chance for me to get lifted, and it was doubtful if the aircrew would be aware of any rifle or machine-gun fire until it was directed at them. Funny how these little problems are never shown until they are staring you in the face.

From the corner of my eye I saw the orange winch-sling arrive from above. A last look around, then I slipped into the sling, waved to the winchman and, holding the rifle tight against me (with thoughts of a friend who, a few weeks before, had his rifle torn from him by branches in a similar situation), my feet left the ground and I was swinging and spinning among the branches on what seemed to be a slow lift to the chopper. On reaching the aircraft I was dragged in by the crewman and quickly slipped out of the sling. Expecting the chopper to get out of there fast, I wondered why it didn't. The crewman looked at me and pointed down. He shouted, but the noise was too great to understand him. Then I realised he had lowered the winch again.

Suddenly all became clear. I was sitting in the chopper with the homing device still switched on and the crew thought there was another man still to come. I sheepishly pulled out the offending instrument and switched it off, signalled to the crewman there were no more to come, then we got out of there, fast.

The LZ was about three hundred yards away and full of a platoon of the Argyll and Sutherland Highlanders. I was glad to see so many friendly faces, but all I could think was, 'Shit! They did it again!' We picked up Pete and were rushed straight to Lundu, where a Twin Pioneer was waiting to get us back to Kuching. A few more minutes and it would have been too dark for the aircraft to take off or land, so we only just made it in time to save ourselves a night at Lundu.

It's quite likely that the crew of the Twin Pin wished we hadn't made it in time. They were not as accustomed to SAS passengers as the chopper crews. We stank to high heaven after not having washed, shaved or cleaned our teeth for two weeks. After sweating gallons, moving through rotting vegetation and stinking swamps, as well as eating curry nearly every night, we were just about humming. My first impression of the aircrew was that they were all a bunch of poofters, but it was only because they had used soap to wash or shower. To us they smelt like the scent counter in a big store.

The RAF Helicopter crews were marvellous in Borneo, as elsewhere. Many of our patrols were lifted from under the noses of the opposition at some considerable risk to the aircrews. In our case, with an expected strong follow-up or interception by enemy forces, no one knew how close to us they were at the time of lift-out.

Having been asked by several people, who knew little or nothing of jungle operations, how we could get back to our LZ in little over one day's march when it had taken three or four days to go the other way, perhaps a little clarification is called for. First of all, the aim was to reach the river and stay there for several days to gain information. Due to the massive reaction we could expect if detected by the enemy close to their supply lines, it was imperative to avoid any contact on the way in, as it would have made it almost impossible to have remained hidden long enough to complete the task; not to mention the difficulties of then getting out again. Secondly, we did not know the extent of the swamps. So we had to try to get through them before finding our way around their perimeter. We also found some very dense jungle vegetation with more than the usual density at ground level.

On the way out we had to move fast to keep ahead of any pursuit. We did little about hiding our tracks, except where we changed direction, as mentioned, and were also able to take advantage of Old Joe's 'good going', which we had purposely avoided on the way in.

Years later I was told the Indonesian officer, whose life was spared by the presence of a lady, was Colonel Moerdani, of the Indonesian Parachute Regiment, who was then commanding the RPKAD, an élite para commando unit in the area. He later led the peace mission which went to Kuala Lumpur and finally wound-up the hostilities in Borneo. Some years after that, Moerdani came to London as a general and somehow Pete and Kevin got to meet him. Their opinion is he's quite a 'good lad'. Worth saving!

The patrol task to the Koemba was typical of the operations carried out by all of 'D' Squadron's patrols on that trip to Borneo. Each patrol's work was made easier, often possible, by work done before by other patrols. Information fed back to our Operations Centre by previous patrols certainly made my job possible. We all looked upon our work as squadron, if not regiment, effort.

There was no brainwashing, or political rubbish involved in making men drive themselves to the point of collapse, risk horrific injuries and a lingering, screaming death in a war thousands of miles from home. We were defending a friendly nation from the military might of its neighbour. We had no doubt about who was invading whom. The Commonwealth infantry units, who took the brunt of the war, were in a thin defensive line, they were in no position to go on the offensive. There is no place like the front line to find out what is going on in a war.

The operations we carried out undoubtedly saved many lives on both sides. In my case, that one ambush on what had been a secure supply line had caused the opposition to pull back troops from the front to guard it. Those troops still had to be supplied and a supply line can only cope with so much. The fact is, there were no more Indonesian attacks across the border in that area after we hit the supply line. It could no longer support enough troops at the sharp end. It had been proved vulnerable. Intelligence reports soon after the Koemba ambush told us the boat had been destroyed, seven Indonesians killed and several wounded. A small price compared to the number killed and wounded on both sides had the enemy's attacks across the border continued in that area.

We also heard that the follow-up force had only followed us for a couple of miles. Whether anything we had done threw them off the track, or they decided it was too dangerous to continue, was never known.

My patrol's efforts were an average minuscule part of the regiment's contribution in Borneo. We knew before we went in, whether we tried or not, whether we lived or died, the squadron, the regiment, the army would go on regardless. Commonwealth forces would achieve their aims in Borneo. So why bother?

In my case, the challenge. I considered myself an expert at my job, an expert with a big 'L' plate. No one knows all about anything much, least of all about my trade. I always knew I would never learn it all but I had to keep learning, keep trying. Operations were like a great chess game to me, with the highest stakes. There was no greater thrill in life than to pit my wits and whatever expertise against a good opposition. Not so much to kill him as to outwit him. The more difficult the job, the greater the odds against me, the better I liked it.

There is no way I can speak for any other member of the regiment when trying to pin down what makes a man drive himself to the limits of danger and endurance. But, if there is one common factor, I would say it is that which is destroyed by putting drill and bullshit above real military training; pride of arms.

16 ADEN PROTECTORATE

'D' Squadron returned to the UK soon after the Koemba job. After a week or two on leave we were back into the mad rush of training and exercises. All those skills which are not kept sharp by spending four months on jungle operations had to be aired and sharpened. There were also various little jobs to be done around the world to keep everyone happy and in touch.

Two courses of instruction came my way about that time. I was trained as an interrogator at an Army Intelligence School. As it turned out, I never had the opportunity to try out their theories for real, but was called upon from time to time afterwards to interrogate 'prisoners' on army exercises.

On the same course there was a man from 'A' Squadron whom I only knew by sight. We were roughly the same size and had similar appetites. We got to know each other during that course, but few people realised we'd met. Sometime later, people were going to put a lot of money on which of us could eat the most steaks in a well-known Hereford steak house. It was one of those inter-squadron rivalry capers, but it fizzled at the last minute. Perhaps they realised we had the situation weighed up and were going to eat to a draw – so we could do it all again.

The other course I did at that time was more interesting and of much more use. I was trained as a forward air controller. This gave me the ability to call on ground-attack fighter aircraft and guide them to targets on the ground. Very interesting. It leaves model aircraft looking a little flat when you can do it with the real thing. There is a special thrill to watching jet fighters attack a target on the ground, especially when you know your words over the radio are guiding them.

In September 1965, 18 Troop went to North Africa for a few weeks. We took our Land Rovers and again got stuck into desert driving and navigation. Every man in the troop became reasonable to good at astro-navigation, using theodolites, star charts and all the other necessary kit. Most of the troop had done quite a bit of desert work before, the two or three who hadn't were in good company to learn quickly, and did so. There was a certain amount of time wasted by the usual clown trouble, but that was to be expected.

It was on that trip I found something I'd never seen before, much to the merriment of some and the horror of others. We were in an area of fairly big sand dunes. The dunes were in long lines like waves. They were about eight hundred yards apart, separated by flat areas of good going. The dunes were about two or three hundred feet high in some places, at least one hundred and fifty feet high at their lowest. We would rush at them at about fifty or sixty mph to have enough speed to reach the top. The top was always soft sand which would stop the vehicle if the driver didn't get the gears right and keep his boot down. Once stopped there was often a lot of hard work to get moving again.

On the far side the vehicle would fall, often two hundred feet to the desert floor, in a great avalanche of sand. This called for slight acceleration and good control to keep the vehicle pointing in the right direction. Paddy had never done this type of desert driving before so I started teaching him. The first two or three dunes he came to he braked when he reached the top. There was a lot of sweat and cursing as we had to dig the truck out and move it on. Then he got the hang of it. We went over two or three more in fine order.

The next one was the one I'd never seen the like of before. I didn't go back up there to check, but I'd swear it had a concrete top! We went up fast, Paddy got the gear right and had his boot right down as we levelled out. The vehicle didn't slow. We accelerated out into space on the far side. About eighty feet down, the tail hit the dune and the nose crashed down. Kevin came over the top of us from the back and did a handstand on the bonnet. Paddy had the sense to accelerate hard to keep control, which stopped Kevin going on over the front and allowed me to pull him back in.

We drove on out, on to the flat, then, just as Kevin had got back to his seat, we hit some soft sand and Paddy braked. This time the Rover stopped dead and Kevin went over us again to land in a heap a few yards in front. There's probably, even now, a patch of blue air somewhere in the Libyan sand sea. Never before or since have I found that type of sand dune with a hard top.

On another occasion, we had been a long way south and were heading back towards our base at El Adem. In the shimmering heat haze ahead we could see a black blob. Then we noticed a couple more off to the left and right.

A few minutes later we reached the one we had seen first. It was a signboard, with its back to us. On the other side, the front of the sign, we found in big black letters the words 'Achtung – Minen!' (Attention – Mines!). Whether or not it was real we couldn't be sure. It certainly looked old enough and we knew there had been vast minefields in the area during the Second World War. The debris of war still littered the desert in many places.

The one thing I always dreaded on operations or exercises was an injury that would put me in the hands of the troop medics. It was no great problem when working in patrols as there was (normally) only one medic to deal with, but when working together as a troop things could get out of hand. This is not to say they were bad at their job, in fact they were very good. The trouble was they were all too bloody keen to put into practice everything they had learnt on their latest medical course. At the first signs of sickness or injury they tended to descend on the victim like a flock of shitehawks, arguing over who would do what and how it should be done. On the other hand, with really bad injuries, there was very little chat and they worked well together as a team.

During that particular trip to Libya one of the men managed to cut his chin rather badly by bringing it into contact with a gun or gun mounting when bouncing over some extra-rough terrain. At the time the troop was all together, or at least within a few miles of each other. I held up progress for the injury to be sorted and told the troop to RV for a brew. Big mistake! Within minutes all patrols were together and four or five medics descended on the chin injury.

Taking advantage of the get-together some of us got our heads together on navigation problems and were engrossed in maps, when I became aware of a low-voiced argument taking place among the medics. Having been alerted by the unusual mutterings, I noticed the odd furtive glance in our direction and, luckily, went to investigate. They had decided who was going to do the stitching job (probably by cutting the cards) and what type of stitches – plain, purl or whatever – but there was now something sinister afoot.

I don't remember how I twigged it, perhaps an unemployed, jealous medic tipped me off, but the bums were going to use very light suturing thread of the type used around lips and eyes – so it would break after a while of rough travel and they could do it all again! Hell! there was enough blood around the place already and they wanted more.

A few dire threats and a few days' travel later, nothing had busted, so I expect they finished up using para cord.

The last couple of weeks of our stay in Libya were taken up by a big military exercise. There were more British tanks in Libya than at any time since the Second World War. Our job was to find the main armoured columns, report their position, numbers and types, then bring in fighter ground-attack aircraft to make mock attacks. This we did with great success. On every attack we called, the RAF claimed a successful strike.

We were not the only people engaged in bringing in air strikes but, according to information received later, we were the only ones achieving any success. The reason, I believe, was our navigation. We knew exactly where we were all the time, day and night. So when we found a target we knew exactly where it was and could bring the aircraft to the right spot and on the right bearing. The RAF squadron who were involved in the exercise were flying Hawker Hunter jet fighters and were impressed enough to send a nice thank-you letter to our base in Hereford after the exercise.

We didn't realise it at the time, but it was the beginning of a good partnership between us and the Hunter squadrons. Early in 1966 the squadron flew out to Aden. Hostilities in that area had been warming up for some time. 'A' Squadron had taken casualties there and although the opposition was rubbish, as is normal in that part of the world, there were a lot of them.

The port of Aden had been nothing much until it became a British colony before the Second World War. The colony was, in fact, only a small area with the sea port being the only item of any importance within its borders. Aden was well placed as a strategic base and as a link on the long British lines of communication to our commitments in the Far East. Its importance was enhanced by the British withdrawal from bases in Egypt. A fairly large city grew up around the port and it became a strong trading centre. Situated at the south-west tip of the Arabian Peninsula, the port saw some of the trade passing between Arabia and Africa, as well as the more obvious sea route between the Red Sea and Indian Ocean, which passed its front door.

The Aden Protectorate, on the other hand, was a wild, mostly desolate area stretching east from Aden almost six hundred miles to the border of Oman. It bulged north into the empty quarter, where its border with Saudi arabia became rather vague. Its population was very sparse and consisted of various Arab tribes who jealously guarded their tribal areas against all comers, mostly other tribes, as no one else had any use for such inhospitable land.

A fairly flat coastal plain surrounded the sea port, but north of that were barren, savage mountains which went north to the border with Yemen and beyond. In an attempt to bring stability to the area, a federation was formed to link the colony with the various tribal sheikdoms between Aden and the Yemen border. As usual, the various politicians over the years made a grand mess of things, leaving an easy target for subversion and armed rebellion.

Yemen, to the north, had had its own left-wing military coup (communist takeover) in 1962 and had obtained the aid of the Egyptian army and air force to quell any opposition and keep the population in line. Egypt was at that time holding hands with the USSR, so the obvious next step was to instigate trouble in Aden, then called the South Arabian Federation.

Our political people made it easy for outsiders to instigate trouble up-country from Aden by causing trouble with the tribes over what was really no more than an old custom. For hundreds of years, the tribes whose land the Aden to Dhala road ran through had collected tolls from anyone (originally camel trains) who passed through. Dhala is a town near the Yemen border and as it grew more and more people had to pay a small sum to the tribes. It was really a protection racket but was of very little consequence compared with the wars made to try to stop it – and the eventual handing over of Aden to the communists.

We, the British, through short-sighted political stupidity made it very easy for agitators to stir up trouble in the whole of the Aden area. The communists stepped in when they saw the unrest. Their long-term aims were to establish a Soviet base in Aden which would dominate north-east Africa – as can be seen from the wars ever since in Somalia and Ethiopia as the communist tentacles spread. As the politicians bog it up, so the army has to suffer. We were due to pull out of Aden in the not-too-distant future so, from the trogs' point of view, it was all a waste of rations. Aden had never been a popular posting for the British Army; so as far as we were concerned the Russians were welcome to their colony, then known as South Yemen.

The media in Britain made much of the army's predicament in the Aden Protectorate, with stories concerning the terrible 'Red Wolves of Radfan' (local tribes armed and aided by the communists). In reality, the problem was in catching up with the so-called Red Wolves. Apart from sniper activities – using rifles, machine-guns, rocket launchers and even mortars on occasion from the safety of darkness or their mountains, they just didn't happen.

'A' Squadron casualties – two killed and some wounded – had all happened when a nine-man patrol in the Red Wolves' area had been caught on the hop by one man falling sick. The patrol had finished up surrounded by about fifty (maybe one hundred) Red Wolves for several hours of daylight, and had fought their way out, to return most of the way back to base on their flat feet. Put the other way around it would have been ridiculous. Nine Red Wolves surrounded by fifty Brits would have been no contest.

When 'D' Squadron arrived we were based in a small engineers' camp at a place called El Milah on the Aden to Dhala road. The road at that point, as on most of its length, was what we called a graded road. It had been levelled by the Engineers using huge grading machines and bulldozers. So it was essentially a dirt road, easy to mine but also easy to repair.

The village itself was quite small, probably less than one hundred small, mud- and rock-built houses, and a small fort – belonging to the local militia, nothing to do with the army. The military camp was a very temporary affair: twenty or thirty tents, a vehicle park, a helicopter pad, all surrounded by a reasonable barbed-wire entanglement perimeter fence. It was well away from the village, being six or seven hundred yards east of the village and road.

The surrounding area was reasonably flat, mostly gravel and sand, but the jagged rock of hills and mountains rose steeply a couple of hundred yards east of the camp and about five hundred yards west of the road across the valley.

It was a very hot place. Very little wind seemed to touch the camp. The heat haze blotted out the floor of the valley most of the day, but there always seemed to be a few 'dust devils', twisting this way and that, somewhere within sight.

The tent occupied by my troop had a small trench in it near the entrance. The trench was only a couple of feet deep, quite narrow, but it had been partly covered over by sandbags resting on several steel barbed-wire pickets. No one in the troop appeared to take any notice of the trench. It was next to my camp-bed. On our first night in El Milah, about midnight, having been in a deep sleep, I suddenly found myself wide awake with the sound of a mortar in my ears. The echoes were going round the local mountains. Where there should have been the sounds of breathing in the tent there was an uncanny silence. Absolute silence; then the unmistakable sound of an incoming mortar shell. The tent erupted with the sound of camp-beds overturning, and I found myself jammed in the entrance to the narrow trench. Jammed because there were two of us. The coward from across the tent had the same idea as me – and must have moved like lightning to arrive at the same time.

The shell exploded nearby and we two were laughing fit to bust, each with our head and shoulders in the narrow, open part of the trench and our backsides raised to the heavens like a couple of ostriches. We stayed there while a few more shells landed, cursing one another for sheer cowardice and finding all the bum reasons we could why the trench should belong to which of us!

I don't think any shells landed within the perimeter of the camp that time. There were no casualties anyway. 18 Troop were a well-experienced troop. Many of us had been in the north Oman campaign and in various other capers around the Middle East. I remember we were the oldest troop in the squadron at the time, possibly the oldest in the regiment, our average age being thirty-two. And that in spite of having a couple of youngsters of twenty among us. The troop didn't give a damn for the so-called opposition and looked upon operations in that area as another training exercise with the added fun of real targets – if we could catch them.

We were doing about five days on each operation but were limited by how much water we could carry. Lying-up during daylight, watching for bands of armed rebels in the mountains, then spending all night in ambush on a likely track. It was a terrible routine. Sleep was practically impossible during the day due to the intense, baking heat and hordes of flies. Then at night, lying in the cool ambush position, sleep tried to overcome us, just when we had to remain alert.

On one of our first jobs we were 'inserted' halfway by road. Every night three or four armoured vehicles rushed down the road, checking for mines, and checking the culverts under the road for explosives. We and another troop were taken on these vehicles then dropped off when they stopped to check a culvert. The other troop was to take the lead as we pushed off into the mountains. Later in the night we were to split and go our own separate ways to our designated areas.

The leading men moved out, I was at the tail end of the column. We hadn't gone twenty yards when the column came to a stop. After a few minutes we moved off again, then, after about two hundred yards, we stopped again for about five minutes. There were a few mutters from the men in front of me, it's no fun standing around with something in excess of one hundred pounds of kit on your back. Again we moved off and again we stopped. The mutterings in 18 Troop were turning to a growl. I went forward and, arriving at the front end, asked what was going on.

The man at the front explained. His troop commander and a troop NCO were going forward to check all was safe before moving the column forward. When I realised who the two were, I could see the funny side. They were the two Mr Magoos of the SAS – how they had ever joined I've no idea. Both wore spectacles but still couldn't see past their noses with any clarity. Their eyesight was legendary in the squadron. The funny side of things, however, was going to do nothing to help my troop. I realised there wasn't one man in the other troop with desert mountain experience. In other words – they hadn't got a clue. Which only goes to show how bog-ups can be made – even in the best-run families.

Movement in that type of country, if you want to cover any reasonable distance, has to be mostly in the wadis. These are dry riverbeds and can vary in width from a few yards to a few miles. They are generally flat with gravel, sand, solid rock or areas of small loose rocks covering a haphazard pattern right across the whole length. Some wadis are thick with camel-thorn trees and various low bushy vegetation; others are almost bare. Most have large areas of completely open ground where it is impossible to walk without being seen perhaps for half a mile or more.

The edge of a wadi can vary considerably but is usually either hills of gravel and rock, solid rock cliffs or jumbles of great rock slabs. Sometimes, obviously, a mixture of everything.

Between the wadis, the hills and mountains rise steeply to jagged, often razor-edge summits. It is always difficult to move along the hilltops – even in daylight. At night it is almost impossible. The only routes through the mountains are the wadis and the occasional tracks or paths linking them.

Operational movements in hostile areas of that type of country are, therefore, hazardous due to the possibilities of ambush. From the high ground in daylight, an enemy can see you approaching at a great distance, giving him plenty of time to organise a hot reception. At night, things are different. Ambushes are difficult to organise because the enemy doesn't know which wadi you will use – or even if you will come at all. Nevertheless, there are an abundance of prime ambush sites in any wadi, so moving through them at night can be dangerous.

The technique used by my troop when pushing through that type of country at night was basically the same as that used in the jungle. Our greatest assets were our long experience and know-how, backed up by very fast reactions and the trained ability to shoot by instinct rather than aim.

The leading scout had to be much further ahead than in jungle work, unless the night was particularly black. Our biggest problem was the possible weight of fire which could be brought down on anyone in the killing zone of an ambush.

The enemy was well equipped with light machine-guns and rocket launchers, and it was likely almost every one in the ambush party would be able to see and aim at the leading scout. The only chance for the leading scout, if he didn't discover the enemy before they fired, was for the men behind him to react very fast and take the weight of fire off him by hitting the ambush party hard and heavy.

As with the jungle, there is no answer to a well-laid ambush once caught in it, but very few ambushes are perfect – especially at night. Enemy ambush tactics in Aden were a completely unknown quantity to us, as no SAS group had ever been caught in one.

Our main tactics for night movement were: travel fast and be ready for anything. That way you had the chance of coming upon the enemy too quickly for him to be at his best.

We wanted to get away from that death-march approach as soon as possible. Suggestions that 18 Troop take over the lead position, or at least put a scout group ahead, were turned down flat.

There was nothing we could do to help the other troop but I had to get 18 Troop on to an operational footing as soon as possible. Checking the map, I found we could get away from this stupid situation at the price of a small detour, so we did, arriving at our appointed place just before first light. Any more delay and we wouldn't have made it. The other troop was caught on low ground the next morning, compromised (seen by the opposition) soon after first light, and had to be lifted out by helicopter soon after. They had made less than halfway to their destination.

In the end, they got the best job. Another troop saw a group of rebels at a distance and our erstwhile companions were put in by chopper to intercept them. They were put in too close to the opposition and had their troop sergeant badly wounded before they could get organised, but put up a good show nevertheless. 18 Troop and another troop were lifted into the area to help out.

The lack of good officers then began to show. When 18 Troop landed we were about eight hundred yards from any action. Well away from everything on low, exposed ground. Useless. I requested on the radio, permission to close the enemy. To my astonishment I was told to keep the troop where it was.

My troop officer, being one of the youngest members of the troop and never having been shot at in his life, lacked experience of such situations. He was a good lad, tried hard, would listen to advice – even act on it sometimes – and I liked him as a man. His main attribute was he had managed to evade the Big Needle (the one used to extract all commonsense from officers) but the patter he was getting in the officers' mess must have conflicted with what he heard at troop level, as he was very dubious of our attitude to the opposition.

The squadron commander was obviously too busy to bother with a troop which had been landed too far away and, knowing his men, would leave them to sort things out for themselves. The troop officer had not been with us long enough to read between the lines and couldn't really be expected to, as this was his first action, so I told him I would take some men on to the higher ground overlooking the troop position as a safeguard.

The opposition had a tendency, in those situations, to converge on any skirmish from all around the surrounding countryside and attempt to pick off any strays or easy targets, in an attempt to take the heat off their own people and also to be able to claim participation in any battle which just might go their way.

Face is a great thing in that part of the world; any opposition in the area who didn't join in, when it was possible, would lose face. The troop was in a prime position to save someone's face, certainly not their own.

There were five of us including Hughie, an American Special Forces sergeant we had lugged around since returning from Borneo. He was quite a good lad, with a dry Southern sense of humour. I wondered what Hughie would have made of it all if he had heard the squadron commander's words to me before we left for Aden. Hughie, as an American serviceman, was not allowed into operational areas during his attachment to SAS. These were American orders from somewhere very high up. He had been with my troop in Libya and during other training, so we knew him well. He was a real soldier and had pleaded to come with us. The squadron commander called me to the office and put me in the picture. I told him there was no problem at troop level. Hughie was welcome to come with us. He considered this for a few seconds, then thumped his desk and said: 'Okay Lofty, be it on your head. If you get as much as one scratch on that bloody Yank – I'll have you skinned and roasted before breakfast.'

Having reached ground high enough to give a view of the area, I realised all was overlooked by a peak about seven hundred yards away which was at the end of the ridge we had climbed. So, while heavy firing continued from where the first troop had landed, we scampered along the ridge to take the peak before the opposition got the same idea – if they hadn't already!

About two hundred yards from the peak we sorted out for the approach. I went forward to check for any problems. Seeing none, I signalled the others to close. Pete and Big Ron were coming up through a gap in the rocks when a burst of machine-gun fire sprayed around them. At a guess, I would say the machine-gunner had stood up and was spraying a panicky burst from the hip, as it went all over the place, some of it hitting the hillside fifty yards or more from Pete and Ron. Nevertheless, as usual in that type of country, it made a hell of a row. The lads obviously weighed it up the same way as I had – panicky and ineffective. They were fine examples of the confidence and experience of the troop. They took no notice, just kept coming. Big Ron's white teeth flashed in a grin on his sunburnt face, then he shouted, 'I thought you said it was safe up there!'

I couldn't see where the machine-gun was and he didn't fire again while we took the peak. It is quite likely he was part of an enemy group who had the same idea as us. If so, we had only just beaten them to it. This was fortunate for the troop which was heavily engaged, as from our position above them they were easily seen, as were the enemy.

Having taken the high ground I tried to whistle up the rest of the troop. Our little clown would have none of it. 'The squadron commander says we stay here – so here we stay!' My remark on the radio that the squadron commander didn't even know where the troop was didn't make any difference. Then the troop must have had a couple of spent rounds go overhead and all I could get out of our troop officer was, 'I'm twenty-one today and being shot at for the first time in my life!' How the rest of the troop, chafing at the bit, would take that gem of news I could only imagine.

We were stuck. With only five of us holding the high ground we could hardly put in a decent attack on the opposition's flank. We had a good view of the skirmish but, as the enemy were about three hundred and fifty yards away and below us, they were too far for effective fire.

Pete managed to chew a couple of them a bit with the machine-gun. One maybe dead, another wounded for sure. But once they realised where we were they made difficult targets, so all firing on our part ceased. Those of us with rifles never fired a shot. We never reckoned to fire at them at ranges in excess of two hundred yards under normal operational conditions.

A chopper brought in a mortar crew. Ranging shells were fired, landing not far from where we could see the opposition. I tried to correct the fire, but although I had the best observation point and the squadron commander knew where we were, the short-sighted troop commander of the heavily engaged troop below us knew better. So I kept quiet – until I realised he was trying to range on us! We watched the opposition disengage and move back to new positions. There were not that many of them. The men in my group wanted to go down there and sort them out.

It was tempting, but we didn't know if there were other enemy close by who would take the hill if we got off it, putting us right in trouble. So I elected to stay put, be a spectator to what looked like developing into a classic bog-up. It was all very frustrating. Especially as we knew the squadron had taken at least one casualty, and, as the chopper taking him out had passed low over us as we moved along the ridge and we could see blood all over the windows, then it looked like a bad one.

Our people were being held back everywhere. Troop sergeants were asking permission to close the enemy from at least two other sides and being told to stay put. I turned to Hughie, and asked, 'Do you get officers like this in your army, Hughie?'

He considered the situation a few seconds. 'Well,' he drawled, 'I reckon we got 'em all right, but we don't seem to get 'em all in one bunch, like this!'

An enemy sniper was putting a shot in every half minute at some of our people away to the right of us. We couldn't see where he was located. The great piles of jumbled rocks echoed every shot so that it rolled on, bounding from rock to cliff for several seconds. We decided to try to draw the sniper's fire so we could locate him easier. As bullets travel faster than sound there is a delay between the sound of the bullet arriving and the sound of the gun which fired it. With a bit of experience the length of delay can be used to calculate the distance to the firer, so making it much easier to locate him. In army training the lesson is called 'crack and thump'. The crack of the bullet and the thump of it being fired. Having been shot at quite considerably over the years the troop was pretty good at judging it. So everyone looked and listened in the right direction while I stood right on the peak of the hill to tempt a shot. I had picked up the odd mutter on occasions which went something like: 'If you think they're such bad shots, Lofty, don't forget that in missing you they might just hit us.' But no one was really worried about this particular opposition. We reckoned that they couldn't hit a barn – even if they were inside and the doors were shut. Hughie was lying on a ledge a few feet down the forward slope just in front of me. I think the wrong sniper fired. The shot hit the ground just below my feet and just in front of Hughie's nose. It had definitely come from a lot closer than it should have, showering us with dirt and gravel.

Deciding he was too close to be given another prime target, we kept our profile a bit lower. No one had been able to pin down his range, except we all agreed it was close. He didn't fire again and we never did find him. With visions of being skinned and roasted, but with no complaints from Hughie, he was told to get his arse to a safer spot.

We were not allowed to close the opposition, and were eventually withdrawn at dusk, having killed a few of them, but not enough in my opinion to compensate for our one badly wounded casualty and two choppers full of bullet holes. The most galling thought was that the survivors would claim a victory – having tangled with the Brits and survived to see them run away at dusk.

I expect the decision to hold us back was more political than military, as usual. The whole scene was typical of the situation in Aden at the time. We were committed to pulling out of Aden in the not-too-distant future. It seemed pointless to put us into operational situations just to build up enemy morale by making us look useless.

Quite regularly our base camp was sniped at or mortared. We could have easily clobbered the culprits if only a few of us had been allowed out to play at night. But no, we had to sit and take it. On one occasion at the open-air cinema show, during a Western movie, we had to put up with real bullets coming through the screen. It was rather a lot to ask.

Every morning, a little lad in the engineer unit drove down to a well the Engineers had sunk, to check all was okay with the pump. The well had been drilled to supply water to the village, it also supplied the temporary camp which was our base.

One morning the brave people shot the young lad, presumably because he was a Brit, alone and an easy target. That summed up their guts perfectly. As far as I know, nothing was done about it.

What we were doing in Aden remained a mystery to all of us. Ours not to reason why.

On one operation, moving at night, I had a fall down a cliff which injured my back and both feet. It was comical really. We were climbing a steepish spur of many-coloured rock. I mistook sand about fifteen feet below for a sandy rock and tried to step up on to it. Luckily the sand was soft where I landed – after a beautiful somersault – but the hundred pounds of kit on my back didn't help. It was really a mistake caused by the old urge to live: being too intent on watching the skyline for the first signs of the expected opposition , I was no more than half aware of where I was putting my feet.

At the time I made light of the injuries, feeling a bloody fool, and carried on for the two or three days left of the operation, then marched the fifteen miles from El Milah to Habilayn, a big military base further north. I can't remember why we had to move base on our flat feet at night, but I do know that by the time I reached Habilayn my back and feet were in agony. I was forced to report to the Doc, who made me rest for a week.

While I was laid up, I had the terrible experience of watching from the tent as my troop rushed to helicopters to go and deal with a problem in the Dhala area. I had never taken a back seat before. The suspense was bad until they returned, all laughing their fool heads off.

The last thing I said to them as they left was, 'Bring him back alive!' They knew I meant our child troop officer. They nearly didn't. He obviously thought he would put on a good show for the troop and insisted on edging along a narrow ledge to put a grenade into a cave, when the experienced heads all around him said there was no need – they could fix it quicker with a rocket. He didn't make it to the cave. He slipped from the ledge and plunged out of sight into the ravine below.

The boys fixed the cave in time-honoured fashion, then peered into the depths below. Someone remarked loudly, 'The bastard's dead!'

A shaky little voice came up from the unseen, 'I'm not dead!'

He had been saved by a tree growing out of the rock face, about forty or fifty feet down. So they fished him out and sent him to hospital with some rather large gashes to be fixed. Poor lad. Another case of boyish enthusiasm not keeping pace with hard experience.

We got a lot of snide remarks from jealous people in the squadron who had bigger problems (with officers, as usual) and couldn't unload them. We laughed all the way to the mountains. About the same time there was a small happening in the Habilayn base which was comical, but very bad for the SAS image.

Our senior Magoo, a captain, was outside the wire fence watching one of our mortar crews get some practice. Someone shouted to him to report to the ops room, where his presence was required urgently. He came at a steady gallop, and the crowd of people from other units who were lining the fence quickly parted, thinking the big hard SAS captain was going to jump the six- or seven-foot-high wire fence. Of course, Magoo saw the gap between the troops but not the wire – and ran straight into it. Luckily for him he wasn't hurt, just bounced back on to his backside.

SAS stocks must have been pretty low on the respectability market after that little episode. The mutterings at troop level are best left to the imagination.

It was in Aden that our previous work with the RAF ground-attack people paid off. By good fortune, the Hunter squadron we had worked with so well in Libya was now based in Aden, and a good working relationship was formed. It allowed the SAS squadrons more freedom of movement, knowing we could rely on men we knew to give us support if things got out of hand.

The Hunter pilots were overjoyed at being able to actually see they were hitting the right targets when we called for them. In some cases, I, for one, would rather have closed the enemy and given them no excuse for being beaten. Nevertheless the Hunters saved a lot of leg work and certainly did their job very well.

Considering the time spent on operations and the exposure to enemy fire, the regiment's casualties in the Aden Protectorate were very light. A lot of the credit for that must go to the RAF and army helicopter pilots, who were brilliant throughout. And the Hunter pilots who never let us down, and were also brilliant at their job.

At one stage the RAF, for reasons best known to themselves, made a rule that no RAF helicopter should land on an undefended LZ. As all our LZs were just places we picked off a map, where we wanted to land, we had to get around the problem by landing three or four men from a small army helicopter just before the Air Force troop carrier landed. That way we could say the LZ was defended.

I often wondered what would happen if we hit trouble with the first men on the ground, before the RAF job had landed. Unintentionally I found out. We were then based in Aden itself. The flight up-country took probably forty-five minutes or more. I was in the small army Scout helicopter giving the pilot directions to find the spot we wanted.

Unknown to me my legs went to sleep. I didn't find out until I leapt from the chopper and tried to run to a firing position about thirty yards away. I hit the ground running and went about ten paces before my legs buckled. I went sprawling among the rocks.

The RAF chopper came in like shit off a shovel and the troop were into fire positions faster than usual. The Air Force pilot had seen me go down like I'd been shot, alerted the troop and came in as fast as he could, thinking we had a problem.

We never doubted the guts and skill of the pilots, but a little reassurance from time to time can go a long way.

Before our troop officer put himself out of action he had another experience he was unlikely to forget. We were in observation positions by day and ambushing at night. A bit squeamish about the seamier side of operational life, he was sharing a *sangar* position with one of the roughest, toughest, absolutely non-squeamish NCOs in the outfit.

The first I knew of the problem the NCO got through to me on the radio and said, 'Lofty, I've got to lay an egg.' We were in well-camouflaged observation positions, unable to move during daylight. The reason I was warned of this 'egg-laying' situation was because it was only just after midday. They didn't want to have the position in a brown fog for the duration of daylight, so wanted to know if I thought it would be safe to dispose of by throwing it into the deep shadow of a gully behind their position. I checked their plan with the binoculars, gave them the okay, then asked, 'Hard boiled or soft?' The voice on the radio came straight back, 'Not bloody boiled at all!' My heart should have gone out to our laddie, but it didn't.

A few minutes later an arm appeared from under the netting, a ration bag swinging, ready to throw. I had been checking the area for any signs of unwelcome company. There were none, so I said, 'Okay, get rid of it.' The bag swung to and fro a couple of times then sailed back behind their position, where it caught in mid-flight on a long springy camel-thorn branch. The branch swung back and emptied the contents of the bag all over one side of the position.

The comments going to and fro on the radio are best left to the imagination. Then our lad was asking if there was any chance of getting out of there before dark. I told him, 'No chance! But thanks for taking away all our flies!' I hadn't heard him swear until then.

17 EUROPE AND BEYOND

After nearly three months on operations in the Aden Protectorate we were glad to get back to the UK for a break. A week or two on leave, then back to the merry-go-round of training. Much of the time the squadron was split into penny packets of maybe a troop, maybe one or two men, spread all over the UK and Europe, doing various jobs.

All seemed disconnected, and they often were disconnected jobs, but the wealth of squadron knowledge and experience was forever growing in many different areas. Mostly we worked on problems related to our particular troop, but there was also a considerable amount of cross-training between troops and, of course, a lot of work common to everyone's requirements.

When there was a reasonable crowd of the squadron in the Hereford base, we often did what were called map exercises. This entailed sitting in a lecture room for sometimes two or three days, working through a fictitious operation in any part of the world and solving problems as they were thrown in.

One particular map exercise was complicated by one of the fictitious patrol members having a broken leg at a crucial stage of the operation. The problem being what to do with him. There were several suggested solutions to the problem. Then the officer running the exercise picked on one of my crew and said, 'Okay, Paddy, if you had a broken leg and I left you in enemy territory with a full water bottle, five days' rations, your rifle and two rounds of ammunition, what would you do?'

Paddy didn't hesitate, 'Sir, as you walked away, I'd put both rounds straight between your shoulders.'

On our last trip to Borneo the war situation had changed considerably. Indonesian incursions into Malaysian territory had been well contained, their supply lines and previously safe areas had been severely punished, not only by SAS operations up to squadron strength, but by companies of infantry.

By the time we arrived a full-scale search for peace was in full swing. Everyone had run out of reasons for continuing hostilities and were looking for an honourable way out. For my part I was glad enough that peace was coming to that part of the world but felt cheated somehow by not being allowed to take up where we had left off. The first month or two after arriving we patrolled the border in case of further attacks. It was the usual hard jungle slog but, with little chance of action, it fell rather flat.

The back injury I had collected in Aden gave me some trouble. There was no problem with marching all day with a pack on my back, but at night the pain prevented real sleep. After a few days' hard work with practically no sleep I began to doubt my ability as a leading scout.

On one operation when we were patrolling a particularly steep border ridge, we had to sleep one night in an area where there simply were no flat spots to sleep on. So we slept on a slope like the roof of a house – and I had a marvellous kip. I slept like a log and felt great in the morning; I had been sleeping in almost a sitting position, with my backside on a tree root to prevent me sliding downhill during the night.

For the rest of that operation I tried every night to find another steep slope to sleep on. I couldn't understand why I couldn't sleep like that again. The patrol suffered these horrific camp-sites with not a little honking. They didn't know why we always seemed to be in the worst possible place to sleep, and couldn't understand the situation, especially as I had previously always insisted on the best possible camp sites.

Years later, at a safe distance in time, Pete told me he had fixed my evening brew to give me that good night's sleep.

I would never take any painkillers on operations, being afraid of getting caught on the hop with my senses dulled, but Pete, being patrol medic, decided a wide-awake leading scout was worth taking a chance for. I'd likely have skinned him alive if I'd known at the time.

Just before leaving Britain we had a couple of new lads join the troop. They were young, very fit, full of mad gallop and go, had just finished a jungle-training stint and were itching to get into the real thing. One came with my patrol and I remember Pete kept rubbing it in that this 'child' was only two years old when I joined the army. Not good for morale!

At thirty-five-years-old I was beginning to wonder at what age I would find all this decidedly hard physical and mental flogging too much to keep up with.

On one operation I did a light-order patrol, taking the brat with me and leaving the others to look after the packs. It was just another job. I didn't rush, just went on as usual and didn't notice any problems. Years later I was told the poor brat had been smashed out of his mind by the time we joined the patrol again, complaining bitterly, 'Does that old bastard ever slow down?'

I often think of that, and several other instances involving so-called 'old men' – some much older, faster and tougher than me – when I hear of international sportsmen, footballers and the like claiming they will be over the hill at twenty-eight or thirty. They must be bloody joking. How can you compare ninety minutes of football, no matter how hard, with day after day flogging through the jungle? Combine that with a trained speed of reaction which you – and the men behind you – have to bet their lives on. Real jungle patrolling, in deep jungle, with every sense at super alert has to be one of the most exhausting ways to earn a living. Seldom are you on flat ground yet you must move quickly and quietly through an indescribable tangle, avoid touching trees which can wave a warning to an enemy, climb over or crawl under a seemingly endless mass of obstacles and still keep your rifle ready for instant use. Almost always on a steep slope (except in swamps), almost always in twilight or darker – even at midday. Try that for a game of football – and don't forget the pack.

The last few weeks of that tour we were banished from mainland Borneo. We were told it was requested by the Indonesians in return for pulling their forces back from the border during the peace negotiations. An island at the mouth of the Kuching River became our base. We arrived there in an assortment of small craft, travelling down the river from Kuching. The island was called Santubong. Our new base was named Merrydown Camp. 'Merrydown' because of the dozens of crates of Merrydown wine which somehow arrived there with the first boat. I'd never heard of the stuff before but suddenly we were almost swimming in it. Perhaps it too was banished from the mainland in the cause of peace.

On the way down river we caught a wild pig which was swimming across. It was quite a big one, fully grown and very active. George did a Tarzan job, diving into the water and wrestling the damn thing into an assault boat (with a little help from his friends). He was welcome, there was no way I would have got into the water with all those snapping, flashing teeth and tusks.

The general idea was curried pork, but we had plenty of fresh rations on the island when we arrived, so we turned the pig loose on the beach. The stupid animal immediately ran into the sea and took off with a very fast piggy stroke for China. It must have been a 'swim-o-maniac'. There was no land in sight out there, so someone went out in a boat, lassoed the beast and towed it back. It was then released in the jungle behind the camp and never seen again.

The island was a few miles long and a couple wide, with a lot of jungle and cliffs rising in the centre. Merrydown Camp consisted of one derelict hut, about forty feet long and twenty feet wide. We used it as a storeroom and cookhouse. Sleeping quarters were anywhere you fancied – in the jungle or on the beach. I opted for the jungle, feeling more at home in the trees than on the sandy beach with the sand flies. Our daily routine became training from early morning until lunchtime. Most afternoons were free for water ski-ing, swimming, surfing or anything we could find to pass the time. We had several boats, masses of ammunition and explosives. There was a lot of demolition training, a lot of boat training, a lot of shooting and, inevitably, a lot of fishing. A few fish were caught, but we must have frightened thousands. Catching fish with explosives is easy in a river, not very productive in the open sea.

Becoming frustrated with the poor results gained with grenades and small explosive charges, the great brains of our fishing industry went along the well-worn path of all humanity. The Big One. The Secret Weapon. The Ultimate Solution.

We lesser mortals watched with interest as they prepared the answer to the fisherman's prayer. A big ammo box, appropriately packed, three separate waterproof initiation sets – each timed for twenty seconds. Nothing was left to chance. It took two men to lift it into the boat, then off they went, out to a spot about two hundred yards from the beach, where they claimed there was more fish than water.

The boat stopped, the great fish-war bomb was prepared, then lowered over the side. When all was just right the switches were pulled. The bomb sank from sight. The boat engine revved – then cut out. Dead!

There was a calm, organised attempt to restart the engine. Followed by a couple of fast panicky attempts. One of the fishermen dived off the front of the boat. Some ungenerous types among us on the beach claimed he was doing a fast crawl before he hit the water. With less than ten seconds to go the engine started and the boat stood on its tail getting out of there, picking up the swimmer – without slowing – on its way (thanks to all that boat training).

The 'Final Solution' lifted a great lump of ocean into froth about fifty yards behind the speeding boat. We on the beach were, by this time, curled up with laughter. It had all been great entertainment, better than *Tom and Jerry*, although very similar. A long search of the area afterwards produced three or four small fish, the final touch to a comic opera.

Living in the jungle one soon becomes accustomed to various types of insect life. Most insects are a known quantity, either painful to tangle with, or no problem. The one insect I was never sure of was the centipede. I was never sure if it could bite, sting, scratch or only mind its own business. There were no real horror stories about centipedes. I had never seen or heard of anyone being bitten, scratched or stung by one. Nevertheless I remember a lot of advice about always brushing them off forwards – never backwards – as their claws will scratch and poison if they are pushed backwards. I never did find out if they were trouble.

In Hong Kong I remember waking one night on Kai-Tak ranges with a large centipede on my stomach. I brushed it off forwards in a panic. It did no harm but people told me I was lucky. During our stay on Santubong I had a very large centipede on me on two different nights. On both occasions they were lying around my throat when I awoke. It beat me how they could get around my throat without awakening me, then bring me wide-awake once they were in position. There are few sensations to rival awakening in the absolute blackness of the jungle night with a nine- or ten-inch centipede for a necklace. In both cases I sat up and brushed the insects off with both hands. I'm sure if any claws had been available I would have been scratched. As was normal when sleeping in the jungle, I had a torch, a *parang* and a rifle within reach. Having snapped on the torch it was touch and go whether I used the *parang* or the rifle when I saw the size of those centipedes.

My ambitions with the water ski-ing scene came to a very sudden stop when I became aware of the local shark population. One big fin at a range of thirty yards and sheer terror kept me on the skis until I hit the beach.

Braver souls continued to enjoy the water sports, but not me. My training went well with my nature when decreeing, 'Only take on a dangerous adversary at a time and place of your own choosing.' I would take on any shark you like at the right time and in the right place. The right time for me would be midday. The right place would be the middle of the Sahara desert.

Malaysia and Indonesia settled their differences over the conference table, so we were flown to Singapore. While languishing in a well-known transit camp on Singapore Island I had occasion to take the troop to a nearby Ordnance Supply Depot, to organise some of the regiment's stores which had arrived by sea from Kuching. We finished our task in a very short time then, as is normal when in those kind of places, had a nose around to see what was going in the way of new kit.

We found a huge pile of strange-looking new boots which, we were told, were to be burnt. I checked with a senior Ordnance type and he told me the boots had been sent out for jungle trials. Having been turned down as useless they had to be destroyed.

Almost every man in the SAS who operated in Borneo had, at his own expense, his army boots modified to suit jungle conditions. Now we found this vast store of boots which had been especially designed for the job. They would have needed no modifications to suit our needs. Our problem had been the issued jungle boots, which wouldn't stand up to our heavy wear and tear long enough to last one fourteen-day operation. So we had used our ordinary army boots, which would last for several operations if necessary.

No amount of talking would get us any of those good boots so, as we had no immediate requirement for them, we left empty-handed. Though there was much speculation in the troop about which clerks at which base job had tested those boots.

British Army equipment for men on field operations was, as usual, lagging behind the times. Our olive green (OG) shirt and slacks for jungle operations were good for the job. The 1944-pattern water bottle, water-bottle carrier and mug were also good. Army mess tins were okay too. The compass pouch was good for the job, and that was it. Our belts fell to bits. Our magazine carriers were locally made in the Far East and in short supply, so that the ones we had were repaired time and time again. Belt pouches were a hodge-podge of allsorts. Spare water-bottle carriers were a favourite type of belt pouch. Issued ammunition pouches had to be modified and strengthened to be of any use.

Almost all our personal equipment was repaired time after time with bits of parachute cord or masking tape. It was often said, if anyone could find a way of destroying para cord and masking tape, the SAS would fall to bits.

The amount of kit we had to carry on our belts went far beyond that which the belts were originally designed for. Depending on the man's job in a patrol, his belt equipment would weigh between twenty and thirty pounds. The kit on our belts gave us the capability of dumping our cumbersome, heavy rucksacks and continuing to operate for some considerable time without them. I always estimated my operational capability after dumping my rucksack as a minimum of two weeks, partly based on the knowledge that in a weakened state, with two machine-gun bullets and at least eighteen small pieces of shrapnel in me, covering roughly twenty miles per night, I had lasted ten days without food and had not been too physically weakened by it.

After that night of cold, pouring rain at Kapoet, we all carried a very lightweight waterproof sheet among our belt kit, as we knew only too well the effects of cold and lack of sleep.

Returning to UK from Singapore we once again had a couple of weeks' leave then were immersed in training, exercises and jobs all over the place.

On one particular exercise in Germany we had a better time than usual. We flew over to BAOR with our vehicles to take part in a big exercise in the Prum area. Most exercises involved a lot of footslogging as well as driving, but on this particular caper we were on wheels for about two weeks. Part of the reason was to test the ability of the Rover Troop to exist as a going concern under as near-to-war conditions as possible.

Apart from the time we spent running ahead of our vehicles at night we were with the vehicles nearly all the time. All our exercise requirements were met, and more besides – enough to keep the brass very happy. At one stage of the exercise we were at a loose end for about twenty-four hours, so I sent messages to the other patrols to RV so we could exchange information on various things and be better organised for the more difficult days ahead. We could only move at night, under cover of darkness, so the RV was fixed centrally in a safe area.

Having reached the RV I decided a small German Gasthaus (pub) in the nearest village would be a better place to congregate, so we checked it out and moved in. The troop's six Land Rovers were parked in the church driveway behind the pub, with two men left to guard them. In the Gasthaus we found only two local customers chatting to the landlord over their drinks at the bar. Much to our surprise, landlord and customers greeted us in good English.

We soon found they had all been prisoners-of-war in UK prison camps. One had been a sailor, the other two were army men. They all knew places and people in Britain and had been back there for holidays several times since the war. There were eighteen of us, counting the two on duty with the Rovers, but those three ex-PoWs in the bar would not hear of us buying a drink while we were there.

When I tried to insist on buying them a drink the landlord threatened to close shop, claiming his hospitality was being insulted. The two customers wanted to go and do guard duty, so our sentries could come in. We drew the line at that one. They understood, of course. As we pulled out, about an hour after arriving, I couldn't help thinking if those good folk received half that hospitality on their holidays in the UK it would be a miracle.

Another training exercise was when the whole squadron did some training in Bavaria with the German Jäger battalions (Mountain Troops). We were given a very hospitable time. Having arrived in Bavaria with only our field-training clothes and equipment, we were somewhat at a loss when, at the end of our stay, the German units and the Mayor of Bad Tolz invited us all to a grand dance and general drinking session. There was no way we could be smartened up for the occasion, so the German troops turned up in their field uniform too – jackboots and all. We had a memorable night in the town hall, which went down well after some good hard training in the mountains. The German troops were very much like us. They trained hard and drank hard. They all seemed to have an excellent sense of humour. A good time was had by all.

During that training, two of us spent a week in the US Army barracks at Bad Tolz, running a signals set-up for an exercise. We were well looked after by the American 10th Special Forces Group and used their PX Club for our coffee breaks. After a couple of days using the PX, we noticed that, every time we walked in, within a minute or so a Lonnie Donegan record would be playing extra loud on the juke-box. It was 'The Battle of New Orleans'. We also noticed it was the same man put the record on every time. Then he would sit smiling innocently at us, clearly enjoying the situation. I decided something had to be done, so every time we used the PX I went straight to the juke-box and put on 'The Battle of New Orleans'.

It didn't take long before curiosity overcame the grinning one. He came over to find out what the angle was. I told him, 'We sent a few good old boys over there all those years ago, it's nice to think they're still remembered.'

I think 'Mr Grin' was thoroughly sick of Lonnie Donegan by the time we left. He'd certainly lost his smile anyway.

18 FRUSTRATIONS IN ADEN

One of my last operational jobs was 'D' Squadron's last tour in the Aden Protectorate. Just as we were ready to leave the UK, Ann was due to have our second child, so I stayed in the UK for two weeks after the squadron left. The birth went as planned and, having organised the home front (well, I had tried anyway!), I flew out to join the troop. When I arrived, they had completed their first operation and were due to go in again that night. One of the troop corporals had been running things in my absence, as we had no clown.

It was mid-April. I had been in the cool climate of Europe for at least two months; the heat in Aden was just getting red hot. Having been flown up-country we had a long, hard approach march that night to reach our designated area. I found myself sweating more than usual and needing to drink rather a lot of water. We were to stay in the area for at least five days. We could only carry five days' water ration and we would get a resupply of water and food on the evening of the fifth day if we were required to stay longer.

As was usual with 18 Troop, we split into two half troops for the observation job during daylight hours. This let us observe a much greater area and also gave us the edge if one group was jumped by an enemy force. Operating in half troops was against orders from above, who kidded themselves about the strength of the opposition. At our level we had no respect at all for the opposition and would have preferred operating in patrols of four. On this particular operation some of the troop were away on other jobs; the troop was only ten-strong with five of us in each half-troop position. At night, we moved into a troop ambush position.

Late on the third day I was down to half a bottle of water, and I realised I was going to be a bit pushed for liquid for the next two days. The position we were in sloped down to the south, so we got the full benefit of the blazing hot sun all day. The rocks all around reflected the heat; it was like a frying pan. By the morning of the fifth day, two others had almost run out of water and I had a raging thirst and only enough water to wet my lips. Mid-afternoon came; all of us were by then almost out of water. No one had taken water from anyone else. We drank only what we carried ourselves.

A signal came to let us know we were to stay put at least one extra day and there would not be a resupply of water until the evening of the sixth day. The sixth day was rough on all of us. To make matters worse, one of the things we were keeping watch on was a well. It was about one thousand metres from our position, near a small village, and in almost constant use. It was thought likely that a rebel force would use the well on their way through the area. We watched it through our binoculars and seeing the water glint and splash just aggravated the thirst.

About one hour before dusk on the sixth day, HQ sent a signal telling us the resupply would not be coming until the seventh day and we were to move that night to a different area. At that point, I cracked. Paddy had been listening on the radio to find out if other troops were in action or in any kind of emergency. All was quiet on the radio; no one had a problem. I sent a message stating if we didn't get a resupply of water within thirty minutes we would go and get some. HQ immediately responded, saying the resupply would come right away – as originally planned. In fact we were on the start line to get our own water when we heard the distant snarl of the Scout helicopter. A chopper never sounded better; water never tasted so good!

Kevin was one of those detached from the troop. He had been sent off, with a vast array of weapons (including a rocket launcher). Kevin carried his armoury in something like a golf bag. He claimed he travelled disguised as a brush salesman – at least, that's what he was described as on his travel documents. We had to put up with a rash of jokes from Pete, whose mind ran riot at the thought of anyone answering their front doorbell and being confronted by Kevin. I must admit, a more unlikely brush salesman would be hard to find.

We were involved with an interesting caper with a company of Marine Commandos in a supposedly very hot area. I think it was in the Wadi Taym, the country of the Red Wolves. The plan was good: the Marines went in first by helicopter and, having taken a good defensive position near a rebel village, sent platoons out in several directions a few thousand metres from the position. We were then flown into the defensive position to get a good look at the area before dark. Just before dusk, we were taken out in the helicopters that were to pick up the Marines from the surrounding countryside.

The chopper carrying my troop landed in a well-hidden little wadi, full of trees and bushes. The Marines moved in from the high ground all around and were lifted out, leaving us to disappear into the rocky landscape. To enemy eyes, the choppers had come and withdrawn the Marines back to the defensive position for the night; they would not expect a small party of Brits to be left behind. The rest of the plot depended on the enemy attacking the Marines in the defensive position – an almost guaranteed next move. Our job was to get among the opposition any way we could and take out as many as possible, any way we wanted.

There was one serious flaw with the whole scene: the village. As usual, we were not allowed near it. Nor were we, or the Marines, allowed to shoot into it, for fear of killing innocent civilians, women and children – fair enough. But why not put the defensive position away from any village? That way there would be no risk to innocent civilians. We could have an unrestricted field of fire. Some of us had honked about it at the original briefing for the operation, but to no avail. The powers that be knew best, however, and we were stuck with the village which was about three hundred metres from the Marine position.

Before last light, we heard on the radio that one of our people had been killed by sniper fire from the village. The troop involved was due to leave the defensive position after dark. They had been waiting in a group and were being briefed on their route past the village when the opposition put a couple of rifle shots into the group – a tempting and unmissable target. Needless to say, the fire was not returned. No one could shoot into the village. The trouble with some of our educated types is they don't realise some bastards have never even heard of cricket. They actually believed the opposition would never attack the side near the village in case our return fire hit innocent civilians. It wasn't the first or last time the British Army's codes of warfare had been used to our disadvantage – nor was it the first time these particular gutless wonders had hidden behind women's skirts. The news of one of our friends being killed put us in good shape to take on anything – in fact, the more the merrier. After last light we moved into a position from where we could quickly get to grips with anyone in our area, then settled down to wait.

About midnight the shit hit the fan. Rockets, mortars and machine-gun fire poured on to the Marines – who were quick to respond with some very heavy kit of their own. It was like the Fifth of November and the Fourth of July combined. Very beautiful to watch when it's not coming your way.

There was no enemy in our area so we moved to our northern boundary and went into ambush positions. It was obvious from the start that the main enemy fire was coming from the village, but the Marines were only able to take on those fire positions away from the village. The Marines' fire with rockets and machine-guns was pretty good as far as I could see. It took good effect where they could put it, whereas the enemy fire was anything but accurate, just heavy.

Every now and then, a couple of the Marines' machine-gunners swept our area with heavy fire causing a fair bit of ducking on our part. Of course, the Marines had no idea where we were; what we did was our problem – they had enough of their own that night. Eventually the opposition packed up and went home. No enemy came near us; no one fired from our area; and we couldn't enter other areas in case of bumping our own people.

Our people, including the Marines, had no serious casualties during the night. We never heard what casualties the enemy had but, from what I saw, I would think they had a few. The Marines were pretty cool with the return fire.

Everyone was lifted out by helicopter during the next morning – we having walked into the defensive position. Another flog around for nothing as far as we were concerned. It was typical of the scene in Aden at the time: some you win, some you lose, mostly it's a non-event. Things were hotting up in Aden itself, though. Terrorists were attacking hard military targets, such as school buses . . . We had men working the back streets and alleyways of the town, and they were getting the odd success. Mostly it was a lot of hard, boring work. I was never involved with that side of the operational scene and had no ambitions in that direction. It was hard enough to get a decent target up-country, without trying to sort one out of the teeming masses in the town.

One of 18 Troop's corporals was pretty good at poker and won a pile of cash in his spare time. Being of a generous nature, he decided to buy the troop a mini-bus so that when we were in Aden we could use it for shopping trips. There was one problem with the mini-bus: the corporal got it at a bargain price because it was a hot vehicle – one which had been well used by one of our teams in Aden and was suspected of being well known to the opposition. In fact, the two bullet holes and several shrapnel holes in the bodywork of the mini-bus more than convinced me it was very well known to the opposition. However, it gave us independent wheels and, as we had all heard about gift horses' mouths, we used it to the full – and very handy it was. As we never moved in Aden without a weapon of some kind, we felt quite safe in the old bus. We were never attacked, anyway.

Talking of mini-buses, we were in the middle of a caper near a village up near the Yemen border when a mini-bus full of people came trundling into the line of fire. A couple of our people rushed out and stopped the vehicle, ordered the occupants out and lined them up, face down, at the side of the road. They were helped into the face-down position by one of the boys putting a burst of automatic fire over their heads.

Imagine our people's surprise when one of the Arabs spoke in a strong Brummie accent, complaining about the noise. It turned out the man had been living in Birmingham for years and had worked at Longbridge car plant. He had come on holiday to visit his people who lived in the nearby village.

I think many of the terrorists in Aden town had got their training from old 'B' movies. They were always making cock-ups which often left them looking stupid. One was about to throw a grenade into a bus full of school children when the driver saw what was going on. He accelerated sharply away, leaving the terrorist struggling to put the pin back in the grenade. Another well-trained crew attacked a small checkpoint. They drove madly through in a car, dropping a grenade from an open window as they went through. Of course, the grenade rolled and bounced along the road, keeping pace with the vehicle and eventually exploding right where they didn't want it. Many good tricks only work on the silver screen.

The operations up-country were often frustrating in the extreme. On one operation in particular, the unbelievably slow reactions of some people were hard to take. We were on a job up near the Yemen border and the whole squadron, plus other troops, were involved, but the job began to fizzle out for lack of opposition. Then one of our helicopters, carrying a couple of squadron people on a recce flight, about three thousand metres away, was shot at by a group of enemy. The opposition used machine-guns and rockets on the chopper, but missed. As we had just about finished the one job, we expected to be lifted out to close and destroy the new enemy group. Nothing happened. There was a lot of waffling going on – but no one was doing anything! We ground our teeth and pleaded for the go-ahead to move on the enemy before they could disappear – all to no avail.

It was one hour and twenty minutes before we landed where the enemy had been. By that time they were long gone; out of play; over the border into Yemen – probably claiming another resounding victory in their great struggle against British Imperialism. It was that sort of thing that made operations in Aden hard going. There was a strong stench of politicians everywhere; I refused to believe our field commanders were that bloody thick. It was as if they couldn't wipe their nose without the go ahead from somewhere.

As usual, the politicians make a hash of something and scream for the Army to get them out of the shit – then meddle with everything the Army does until they make the poor bloody trogs look as stupid as politicians. You don't have to look any further than Northern Ireland to see a prime example of an army being treated like a pack of foxhounds – forever held on a short political leash! The politicians never seem to realise it costs men's lives (usually) to show the enemy who is boss. If you don't do it quickly – and keep proving it – the enemy will draw courage from your hesitation, bungling or whatever. Any courage on his part will inevitably lead to harder conflict at trog level, with inevitable heavier casualties on both sides.

The Aden Protectorate was about to be vacated by the British, so really any military operations would only succeed in building up the morale of the anti-British element in the end. They would be the heroes who had driven us out. The whole scene seemed pointless. When 'D' Squadron left Aden for the last time, 18 Troop stayed behind to carry out a long-distance Rover Troop operation. Our vehicles had been sent out to Aden along with all the equipment and weapons for the operation. I believe the troop was probably the best and most experienced Rover Troop in the regiment at that time. The planned operation was decidedly dicey in many respects, but we trained hard and were very confident in our ability to pull off a successful job. In the event, after a lot of hard work, I had to cancel the operation due to half the troop being rushed off on an emergency (to nursemaid some Foreign Office people somewhere). Some people wanted me to do the operation using spare clerks, storemen and such, who belonged to the squadron that had come out to relieve us. We waited a couple of weeks, hoping to get our men back. When they didn't return, and no one could tell us when they would, I arranged to get the remainder flown back to the UK as soon as possible. It was another instance of the regiment being overstretched. Everyone, military and otherwise, realised the potential of well-trained Special Forces. At that time the regiment had four squadrons and could likely have found plenty of work for twice that number.

A few more mostly non-eventful operations and exercises came my way before leaving 22 SAS to spend the last years of my army service with 23 SAS, one of the reserve units.

It was not until the last year or so of my service that I realised I had survived. Until then I had never thought of being out of the army.

Somehow, deep down, I knew survival was not for me, knew that one day the luck would run out. Those feelings probably went back to my first experience of war, in Korea, where it was obvious my chosen profession did not offer a lot of future, and the thoughts of another twenty-odd years with that sort of thing, possibly every couple of years, well, all thoughts of life after the army vanished. 'He who lives by the sword is very apt to die by the sword.' Hopefully it would be quick. I would not be another obscene mess, writhing in the mud, coughing my life out.

As a British soldier I was never trained or encouraged to treat lightly the lives of other people. I don't think I was brought up or trained differently to anyone else of my generation. Over the years in the mass media, there has been a lot of talk and tales of horror associated with people's reactions to the act of taking human life. Perhaps there is something wrong with me: I have never experienced any regret, remorse or any other emotion I can honestly remember – except relief. Relief that it wasn't me lying there dead or screaming. Relief I had acted first and fast. Relief my weapons had worked as they should. Relief I had not betrayed my comrades or myself with any fumbling, mental or physical.

In my experience there is no time for worrying after having shot someone, as your biggest problem is who will shoot next. Shooting one or two enemy, if you are lucky to be fast enough, can often bring a mass of hostile fire from unexpected quarters.

The last shots I fired on operations with the British Army were intended to warn, not to kill. A woman was walking along a path where she should not have been. We knew she was walking towards almost certain death. Had it been a man he would have been taken out with the machine-gun, but it had to be a bloody stupid woman. So I elected to warn her with a well-placed rifle bullet.

The range was rather long but I managed to bounce the shot off the track about three feet in front of her. She stopped, didn't look left or right, just stood! Then she started to walk very slowly forward again. Another shot, which bounced off just in front of her feet. Again she stopped. Just stood, stupidly facing the way she wanted to go. After about half a minute she very slowly shuffled forward again. The third shot must have almost hit her feet. She stood perfectly still. Couldn't she take a hint? What a player – though God help any poor son-in-law she might have. A few feet further and she would be out of sight. I decided that if she went on she could damn well take her chances. Putting down the rifle I watched her with binoculars.

She stood there a long time, perfectly still, facing resolutely where she wanted to go, not so much as a half glance to either side. After several minutes she shuffled very slowly out of sight. I cursed the sheer stubbornness of the woman, while we listened for the sound of her demise. Then someone said: 'Hey, look.' We all looked. There was the woman, walking slowly back along the track, leading a huge brown bull, which had obviously been tethered out to graze. If she had as much as half-raised two fingers in my direction I think I would have shot her clean between the eyes.

She may have had a perfectly innocent reason to be there, but in that land of cheap life, well, she had more bloody nerve than I have.

Some people may pause to wonder what kind of man, or animal, is produced by a background like mine. Most of the gory details and some best-forgotten experiences have been left out of my writing. Thinking back over the years, in common with most soldiers, I don't know if I have ever killed another human being. I have seen a lot of men go down, but whether it was my actions which killed them is open to question. Not many soldiers have the benefit (or otherwise) of medical or forensic experts to certify death or its causes in an operational situation. As far as I am concerned, I did what I had to – no more, no less.

If, on any operation, my mother had been looking over my shoulder, she may have been considerably shocked, but never ashamed.

GLOSSARY

BLIGHTY	UK
BLITZKRIEG	Lightning War (German)
BMH	British Military Hospital
BULLSHIT	That which supposedly baffles brains,
	a) By actions such as polishing, painting, scrubbing, pressing and 'smartening' to a point far in excess of clean and tidy
	b) By talking a load of crap
CHOPPER	Helicopter
CQB	Close Quarter Battle
DEMOB	Demobilisation (Discharge from the Army)
DZ	Drop Zone, landing area for troops, equipment or supplies dropped by parachute
ETA	Estimated Time of Arrival
FAC	Forward Air Controller
FGA	Fighter Ground Attack (Aircraft)
GURKHA	Soldier from Nepal, serves in their own distinctive units in the British and Indian Armies
HE	High Explosive
IA	Immediate Action
LAST LIGHT	Dusk, Sundown
LZ	Landing Zone, for Helicopters
NAAFI	Navy, Army and Air Forces Institute (British Forces Canteen and Shop)
NATO	North Atlantic Treaty Organisation
PARANG	Machete, long jungle knife
POW	Prisoner of War
PT	Physical Training
RECCE	Reconnaissance
RSM	Regimental Sergeant Major
RTU	Return to Unit
RV	Rendezvous, meeting place
SANGAR	Rock walled position where digging not possible
WADI	Any type or size of natural water course (Arabic)

WEAPONS

BREN
Type of Light Machine Gun
Cal: .303inch(1950s) 7.62mm (1960s)
30 round box magazine

BROWNING
(Thirty Calibre)
American, Medium Machine Gun
Cal: .30inch. Belt feed

BROWNING
(Fifty Calibre)
(Point Five)
American Heavy Machine Gun
Cal: .50inch. Belt Feed

'BURP GUN'
a) Soviet PPSh 1941G. Sub machine gun
Cal: 7.62mm. 71 round drum or 35 round box magazines
b) Chinese. Type 50. Sub machine gun
Cal: 7.62mm. 35 round box magazine
(derivative of Soviet PPSh 1941G)

GPMG
(General Purpose Machine Gun) British Infantry Section Weapon. Cal: 7.62. Belt Feed

GRENADES
NO 36 British. HE shrapnel grenade
NO 80 British. Phosphorous grenade
NO 94 British. Anti-Tank grenade fired from rifle using (Energa) ballistite cartridges. Approx. range 150 yards

'STICK'
Chinese. Blast grenade with approx. 7 inch or 8 inch wooden handle

HEAVY MACHINE GUN
Used by Chinese. Probably Soviet. DshK
M1938 with wheels and shield. Cal: 12.7mm. Belt Feed

LMG
Light Machine Gun

MMG
Medium Machine Gun

MORTARS
2 inch. British. Infantry platoon or section weapon. Fires smoke or H.E. shells or parachute illuminating flares. Range 475 metres
3 inch. British. Infantry Battalion Support weapon. Fires smoke/phosphorous or HE shells or parachute illuminating flares. Range approx. 2560 metres
81mm. Infantry Battalion Support weapon. Fires smoke/phosphorous or HE shells or parachute illuminating flares. Range approx. 3200 metres. (Range has since been considerably improved)

	4.2 inch. British Artillery weapon. Fires smoke/phosphorous or HE shells or parachute illuminating flares. Range approx. 3750 metres
PISTOL	Browning. Semi automatic hand gun. Cal: 9mm. 13 round box magazine
PISTOL SIGNAL	(Varey pistol) 1 inch. Single shot. Red, green or illuminating cartridges
RIFLES	*Armalite*. AR15. American light automatic rifle. Cal:5.56mm. (.223inch) 20 round box magazine. Used by SAS in Aden and Borneo, also by some Indonesian Forces

FN Rifle. Belgian (Fabrique Nationale) Automatic

Rifle. Cal:7.62mm. 20 round box magazine. Replaced bolt action Lee Enfield in British Army approx. 1957 until British version (SLR) became available in early 1960s

Kalashnikov. AK47. Soviet Assault Rifle. Automatic

Cal:7.62mm. 30 round box magazine. Some used by communist infiltrators in Aden and other parts of Middle East and Africa

Lee Enfield No.4. Bolt Action.

Cal:303. 10 Round box magazine. British Army rifle through WWII, the Korean War and up to approx. 1957

Martini-Henry. Single shot, lever action.

Cal: .45 inch. British Army weapon of the 1870s. Was still in use in various parts of the Middle East and Asia until well after mid 20th Century

Simenov-M1936(AVS). Soviet. Automatic rifle with folding bayonet. Cal: 7.62mm. 10 round box magazine. Used by some Chinese forces in Korea. Seen carried by Division or Regiment Headquarters personnel of 63rd Assault Army

SLR (Self Loading Rifle) British Army semi automatic rifle from approx. 1961. Cal: 7.62mm. 20 round box magazine. British derivative of Belgian FN. (FN adopted as standard NATO weapon)

3.5 INCH ROCKET LAUNCHER	Anti Tank weapon. Improved and heavier version of WWII Bazooka. Range 900 metres
SMG	Sub Machine Gun
STEN	(Sten Gun) British. Sub Machine Gun. Cal: 9mm. 32 round box magazine
TOMMY GUN (TSMG)	American. Thompson Sub Machine Gun Cal: .45 inch. 30 round box magazine or 50 round drum magazine
VICKERS MMG	British. Medium Machine Gun. Infantry Battalion Support weapon. Cal: 303 inch. Belt feed. Water cooled. (Superseded by GPMG in early 1960s)

ARMED FORCES INDEX

Australia
Australian Army 13, 238, 241
  Military Police 210

Bermuda
Bermuda Militia Artillery (BMA) 208–210

Britain
British Army 74, 188, 189, 223, 259, 264, 278, 299, 301
  Argyll and Sutherland Highlanders 271
  Army Intelligence School 274
  Centurion tanks 19
  1st Gloucestershire Regiment (Glosters) 11, 17, 19, 33, 34, 112–13,
  119, 122, 119, 120, 122, 127, 129
        'A' Company 23, 24, 31
        'B' Company 17, 18, 30–32, 34, 38, 40, 56
              4 Platoon 27, 28
              5 Platoon 27
              6 Platoon 18, 27, 28
        'C' Company 23
        'D' Company 24, 71
        Gloster Meteor reconnaissance jets 152, 176
  Gurkhas 130, 211, 222, 238, 239
  Military Police 192
  Royal Army Service Corps
        55 Company 159
  Royal Artillery
        8th Royal Irish Hussars 19
        45th Field Regiment 19
        170th Mortar Battery 19, 41
  Royal Engineers 146, 201, 285–6
  Royal Northumberland Fusiliers (RNF) 18, 19, 54, 55
  Royal Ulster Rifles (RUR) 19, 54, 70–71
  Scots Guards 237, 250

RPKAD [Resemen Para Kommando Angaton Darat] 272

Japan
Japan Reinforcement and Holding Unit (JRHU) 13, 14

New Zealand
SAS 250

Oman
Company of Omani Northern Frontier Regiment (NFR) 161, 163, 168, 169
Sultan's Armed Forces 146

United Nations (UN)
United Nations 15–16, 56
    UN Command 116
    UN press 110

United States of America
American 10th Special Forces Group 244, 283, 296
Clarke Field, Manila 113
F80 Shooting Stars 27, 32, 54
Military Police 16–17
US Air Force 27, 93, 209
    bases 113
US 1st Marine Division 56
US Infantry 109
US Marine Corps 129, 209
US Navy 209
US Rangers 209
US 3rd Division 56

INDEX

Aberdare forest 196
Aberdare mountains 194
aborigines 139
Aden 205, 277, 278, 281, 285, 287, 300
Aden Protectorate 220, 277, 278, 287, 297, 302
Adlington, Captain 145, 149
aerial combat 92
Africa 277
age 290–91
'Aggis' 90
air raids 87–8, 93, 107
air strikes 276–7
air war 92
air-drops 140, 159–60, 167, 168, 180, 183, 185
air-photography 228–9
Akbat 161, 163–7, 168, 169
Akhbat el Zhufar *see* Akbat
Aldershot 129
Alex 170, 173, 175, 178, 224–5, 228, 229, 237, 242
Allied forces 222
ambushes 53–4, 130, 191, 234, 236, 249, 258, 259, 261–3, 267, 281, 288
American troops 15, 16
ammunition 22, 27, 50, 55, 148, 154, 159, 175, 185, 294
Amsterdam International Airport 214
animal tracks 225, 227
Ann [Large] 12–13, 78, 115, 116, 121–2, 124, 127, 128, 139, 140, 144, 171, 190, 194, 196
anti-aircraft-fire 87–8
antibiotics 246
Anzio landings 238
apes 227
AR16 (Armalite) 253
Arabia 277
Army Certificate of Education First Class 144–5

Army Education Centres 121, 145
Army Medical Centre 128
artillery-fire 32
attap 134, 136, 137, 196, 226
Austin K9 trucks 201

bacteriological warfare 93–4
Bad Tolz 295
badges 21, 57
'Baffs' (British Armed Forces Special Vouchers) 65
bagpipes 169
Bahrain 167, 202, 205
Bait-el-Falage 146, 148, 169, 187
Baker, George, Corporal 45–6, 47–8, 57, 63
BAOR 294
Baraimi Oasis 204
barbed wire 68
battle camps, Hara-Mura 13
Bavaria 295
bayonet warfare 28
BBC Overseas Service 101
bears 211
beatings 97
beer 169
belts 294
'Bergen Pie' 244
Bergen rucksacks 175, 135, 216, 244
beriberi 72, 76–7
Bermuda 208–210
Bermuda Rifles 208
Bert [Perkins] 125, 127, 238–40, 241
Biffo 139
Big Doc 200, 207
big game 196
Big Needle 282
Big Ron 283
bikini 131–2
Bill 176
binoculars 217